Tom Fort read ~~Engli~~ .... ~~w~~ork as a
reporter on a newspap~~er~~ .... ~~w~~orked for over 20
years. He is married with five children and is the auth~~or of six~~ other books: *Under
the Weather* published by Century and *The Grass is Greener*, *The Far From Compleat Angler*
and *The Book of Eels*, published by HarperCollins.

Praise for *Downstream*

'In *Downstream*, Fort sets out by punt to explore what he describes, in a character-
istically lovely phrase, as "the concealed dimension" of water. Fort is a quintes-
sentially English guide.'
Observer

'For the river-lover, who likes to pause and look over bridges to see what is going
on beneath (and alongside), and who likes to think of the past and the present
continually flowing into the future, this book will be a delight.'
The Spectator

'The story of the river that bubbles up throughout the narrative is told with the
quiet authority of one who knows about the hidden flow beneath the riverbed. Well-
referenced scholarship lies behind Fort's lively accounts of the history of this part of
Middle ~~E~~ngland. Erudition with a light touch, cleverly interwoven with stories of
pints d~~runk~~, clean sheets and greasy breakfasts appreciated and characters met;
what ~~fun it i~~s and how impressive is the learning that lies behind it.'
TLS

'Fort b~~rings~~ an almost lyrical beauty on rivers both actual and archetypal.'
Daily Telegraph

'It is dif~~ficult n~~ot to like Fort, if not simply for his encyclopaedic knowledge of
all thing~~s fluvi~~al then for his wit and irony too. Once engaged it is difficult to
escape. He ~~has~~ stirred a latent fondness for oxbow lakes and wandering streams.'
Literary Review

'A well ~~obse~~rved travelogue. [Fort] has cleverly constructed the book to reflect the
river's characteristics. Tom has written a charming book that cannot fail to give
pleasure.'
The Oldie

'Fort is an amiable companion, whether investigating the river's pubs, observing
its wildlife or finding an enviable freedom camped out on its banks.' *FT Magazine*

'*Downstream* is a double delight; a celebration of the River Trent itself and a
humourous exploration of its historical, geographical, industrial and cultural
curiosities.'
Nottingham Evening Post

Praise for *Under the Weather*

'Delightful book . . . every one of us should have a copy of this book on the shelf.'
Daily Express

'Reassuringly warming, the pleasure of this book is in the rambling flavour of his
travels. He writes beautifully.'
Daily Telegraph

'Readers will not fail to be awestruck. Fort is a robust writer.'
Daily Mail

'Fascinating, beautifully written. A rich concoction of different genres.'
New Scientist

Also available by Tom Fort

The Grass is Greener
The Far From Compleat Angler
The Book of Eels
Under the Weather

TOM FORT

DOWNSTREAM

arrow books

Published in the United Kingdom by Arrow Books in 2009

1 3 5 7 9 10 8 6 4 2

Copyright © Tom Fort, 2008

First published in Great Britain in 2008 by
Century
Arrow Books
Random House, 20 Vauxhall Bridge Road,
London SW1V 2SA

www.rbooks.co.uk

Addresses for companies within The Random House Group Limited can be found at:
www.randomhouse.co.uk/offices.htm

The Random House Group Limited Reg. No. 954009

A CIP catalogue record for this book
is available from the British Library

ISBN 9780099505662

The Random House Group Limited supports The Forest Stewardship
Council (FSC), the leading international forest certification organisation. All our
titles that are printed on Greenpeace approved FSC certified paper carry the FSC logo.
Our paper procurement policy can be found at www.rbooks.co.uk/environment

Mixed Sources
Product group from well-managed
forests and other controlled sources
www.fsc.org Cert no. TT-COC-2139
© 1996 Forest Stewardship Council
FSC

Typeset in Spectrum MT by Palimpsest Book Production Limited,
Grangemouth, Stirlingshire
Printed and bound in Great Britain by
CPI Bookmarque, Croydon CR0 4TD

To the memory of my father Richard Fort

Picture Credits

I am grateful to the following people and organisations for their help in identifying photographs, postcards and prints in their collections, and for permission to reproduce them:

John S. Booth of Berryhill, Stoke-on-Trent

Brian Lund of Reflections of a Bygone Age, Nottingham

Andy Nicholson of www.nottshistory.org.uk

Glyn Hughes, Senior Collections Officer, Newark Millgate Museum

Chris Copp, Museums Officer, Staffordshire Arts and Museum Service

Ros Boyce, Illustrations Indexer (retired), Lincolnshire Archives, Lincolnshire County Council

Katey Goodwin, The Potteries Museum and Art Gallery, Stoke-on-Trent

I also owe a great deal to David Hall of Blueskyimages, who worked wonders with my own mediocre snaps.

Contents

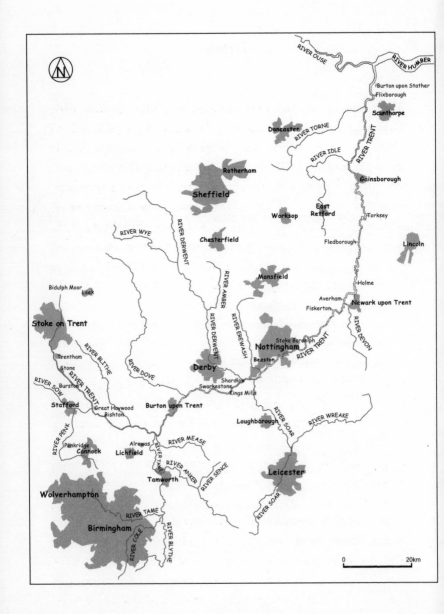

Preface

There should be a word for someone who takes a particularly intense pleasure in the company of rivers. Fluviophile is both clumsy and pretentious. River-lover is better, but still slightly awkward, and you have to be precise about the nature of the 'love'. It's not enough to sigh 'Isn't that pretty?' or 'Isn't that picturesque?' then wax lyrical about the weeping willows or the ducks and coots and swans or the painted narrowboats and slender skiffs. The river-lover's clear eye is on the water itself: its motions and texture, its colour and clarity, the way it reflects the sky in silver bands and the dark humps of the trees, the way its surface is marbled by the interaction with the stream bed.

The river-lover sees more than just fresh water on the move. There is awareness of past, present and future – that the water was somewhere else the day before and will be a different somewhere else the next day; that it connects places and people across space and time; that it is both of itself, self-contained, and integral to its landscape. There is a curiosity about its story, how it was seen, how it impinged on lives, how it was used, which armies halted at its banks, which poets strolled through its meadows, which traders put their trust in it to take them where the business was.

You can tell a river-lover. They cannot help but pause on a bridge to investigate what lies beneath, or at least slow down to steal a look. The river-lover must try to see below the surface, to apprehend the concealed dimension. Down there are strands of weed swinging in the current, and slopes of gravel and flats of mud, boulders, stones, blackened logs, bones, wheel hubs, trolleys, coins, rings, rubbish and treasure. As a young reporter I once covered an inquest solemnly appointed to decide whether a jewel-encrusted gold sword that had once belonged to Nelson and had ended up at the bottom of the Thames near Windsor was the property of the Crown or the diver who freed it from the silt.

Down there, burrowed into the same silt, are the nymphs of the damselfly and the mayfly that will take wing the next summer or the summer after that. The impregnable invader, the signal crayfish, creeps across the bed snaffling insects, eggs, corpses, anything it can get its outsize claws on. Stuck to the underside of the stones are the cases of the caddis larvae, like minute fragments of twig, and the webs of the buffalo gnat. Stoneflies dart through the dark interstices. Freshwater shrimp and snails and leeches cling to the weed. The order of fishes observe their places, from the minnow the length of a thumb-joint to the pike as long as your leg. Shoals of fry shimmer in the shallows. Occasionally there is a swirl at the surface as a predator strikes and its prey scatters, but mostly the underwater world conducts its business out of sight and silently, and you must inquire quite persistently to get an inkling of what is going on.

These are just some of the ways in which the particular affinity for moving water shows itself. There is as well – dare I say it? – a metaphysical aspect to all of this, even spiritual: a sense of the sacredness of rivers as givers of life. To me the sacredness is free from religious association, though some have seen it and see it differently, the hand of God at work. The spectacle gives me cause to wonder and inspires a kind of reverence. But maybe it's best to leave the expression to the poets, for fear of courting charges of mystical incoherence. Ted Hughes was abnormally sensitive to the wonders and mysteries of rivers, as you might expect from a poet who spent so much of his time beside and in them pursuing fish. He wrote a whole volume of verse about rivers and included in it a poem called 'River' that attempts to distil some of the transcendent elements. It ends with the lines

> It is a god, and inviolable.
> Immortal. And will wash itself of all deaths.[1]

I suppose I may have sensed something of that even as a boy, when I was first allowed to wander a stream on my own. Certainly I have always been conscious of the unique nature of moving water being special to me, even though I have not really struggled, as Hughes did, to find words to express the feeling. Over the years I have written a great deal about rivers, but always from the perspective of the angler, always about fish and trying to catch them and what that pursuit meant to me. Gradually I came to want to try something different: to find a way to write about the river, all rivers. I wanted to discover how they took shape, how they helped make and define the landscape. I wanted to understand how they work in the sense of their physics. I wanted to describe how they had nourished the world for us, provided routes for trade and conquest, given us energy. I wanted to celebrate the ways in which they stirred spirits and set imaginations alight; to learn how they were worshipped and reverenced, and then abused and overlooked.

It was obvious I had to do more than just study the subject. I needed to experience it, to be on the water and of the water. I needed to make a river journey. The question was: which river?

It had to be a big river, inasmuch as we have any. It had to have a certain stature, or at least to have figured in the scheme of things. The nearest to me, and the one most familiar to me, is the Thames. But the Thames bibliography fills a dozen shelves. I have read quite a number of those books and I seriously doubted if I could do anything very different from those who had gone before. I thought of the Severn, second to the Thames in length. But I have long admired Brian Waters' two books about it – *Severn Stream* and *Severn*

Tide[2] – and did not care to stand against him. I thought briefly about the Wye. But Robert Gibbings treated the Wye, as he did the Thames;[3] and anyway I had reservations about navigating its brawling upper reaches.

Then I thought about the deeply unfashionable Trent. There have been books about it, but not many, and no one had floated down it and written about it. Indeed, until recent times it was so polluted that you could not have done so without the risk of poisoning yourself. I knew almost nothing about the Trent, except that it had come back to life after the best part of two centuries as one of the filthiest rivers in England. I had a very imprecise idea of where it rose, and was hardly acquainted with the counties it flowed through (for your information: Staffordshire, Derbyshire, Nottinghamshire and Lincolnshire, and beside a little bit of Leicestershire). My ignorance was widely shared. 'The Trent, eh?' a river-loving friend would say quizzically. 'Where is that, exactly?' Or: 'Where does that come out?' 'The Humber,' I would say. 'Ah, the Humber,' he would respond, heaping on the quizzicality, as if I had said the Gulf of Aqaba or the Bay of Bengal.

I went to take a look. At 170 miles long, it carries a lot of England with it. It cuts down through Stoke, then across, then up, through the heart of the Midlands towards – but not to – the sea. When I looked in the old books I found it had a good deal of history attached to it, yet today it is remarkably hidden away. Only three significant centres of population are gathered around it – Stoke, where it is only a brook, Burton and Newark (I tend to discount Gainsborough, which is to one side, and Nottingham, which historically did its best to keep its distance from the river, until it became too big). There are a few villages on the Trent but not many, and strikingly few fine houses and gardens. For the most part it flows through peaceful, unexceptional farmland, going its discreet way, its great days long departed.

But it did have its time. Early settlements were drawn to it, invaders thrust up and down it. Conquering and defending armies paused at its banks and splashed across its fords. It served to mark a division in the land, between the part worth troubling with and the part beyond the pale; and later between Royalist and Parliamentarian spheres of influence; and later still between north and south.

For centuries the Trent was one of the most important trade routes in England, until the railways and then the roads put it out of business. Even thirty years ago there was constant barge traffic as high as Nottingham, coasters plied the wide waters from Gainsborough down to the Humber, and the jetties and wharves bustled. Almost all of that commercial life has gone, until you get far down on the tidal stretch where big cargo ships can operate. A handful of barges still carry sand and gravel from workings below Newark downriver and up the Ouse to Wakefield, but they are kept afloat by subsidy. Mostly the river is quiet and peaceful, but it is the peace of abandonment, which always has a sadness about it. The disused jetties stand as dumb witnesses to the decline. The timbers rot and the piles rust and blackened ropes rise from mud-encrusted tyres beside metal ladders that have not felt a footstep in years.

So there was a story to be told and a journey to be made. The practicalities were much less daunting than they would have been on another river. The top section was too small to put a boat on, but that would have applied anywhere. From a few miles above Nottingham down to the Humber, the Trent is an official Navigation. In between – from Trentham, where I thought I could get afloat, to Sawley, where the Navigation begins – was the unknown, unexplored river. I was well aware that, legally, I had no right to be on it. But I reckoned that the chances of encountering riparian owners, and of any of them wishing or being able to detain me, were small. From what I could gather from random checks and talking to informed sources, the water itself should be manageable as long as it was not in flood.

All I needed was the boat. So I obtained the boat. And I did the journey, mostly in the boat, some of it by other means of loco-motion, as explained later. And it was wonderful, most of it, in ways I was not prepared for. I paddled and rowed and waded and heaved and strained and got fitter and leaner than I have been in decades. I slept beneath the stars (separated from them by tent) and cooked for myself at the water's edge. I was alone, which I liked very much indeed, and was entirely dependent on myself, which gave me a little trouble at first but which I also came to like. I explored a path through an England I did not know, and I think I got as close as I could to the river, and that was the best part of all.

There's no point in my apologising for the fluvial character of the story. It meanders, I know that. It has to. That's what rivers do, and it would be a very strange river book that kept in a straight line. Some of the more extreme meanders resulted in oxbow lakes and failed to survive the critical scrutiny of my editors at Century, Mark Booth and Charlotte Haycock. But they were generally right, and proved rather more indulgent towards the windings than I had feared. I am grateful to them.

I am indebted to the following: my indefatigable agent, Caroline Dawnay; Jon Beer, who made the boat for me and gave me more pleasure from it than I could have dreamed of; Andy Nicholson, who supplied the map and a number of illustrations, displayed endless patience with a wide range of demands, and runs the invaluable website www.nottshistory.org.uk; Tim Jacklin of the Environment Agency, who advised me wisely and provided much obscure information about fish; Brian Bettison of Stoke Bardolph, who kept me afloat; my friend Steve Taylor, who pointed out things that needed pointing out; my wife Helen, who kept me going in many ways. I am also grateful to my father and mother-in-law, Michael and Bar, for helping me get on my way; to Trentham Leisure for letting me launch the *Otter* and to Burton Leander Rowing club for letting the *Otter* stay.

Chapter 1

'I go on for ever'

Explorer

One hot afternoon in the middle of June 2005, I said goodbye to my wife and our two daughters, pulled my straw hat down to shield my face from the sun, and climbed over a stile into a field to search for the source of our third longest river, the Trent. Above me, straggling along the ridge, was the village of Biddulph Moor. To the west, topped by chocolate-coloured stone outcrops, rose Robin Hill. Beyond that, invisible, was the town of Biddulph, where coal had been dug for a thousand years, but no more.

The village has some charm, of a slightly dour kind. But Biddulph struck me as more grim than dour, and in the middle of one of its grim housing estates there was a pub called the Top o' the Trent, a stark block of brick roofed by green tiles. I had heard of it when — far away in leafy Oxfordshire — I started planning this journey. I thought at once: that is where I must begin, there can be no other starting point. I changed my mind when I saw it. It was bolted and barred, deserted and lightless; which was as well, since it was a dispiriting place and, apart from its whimsical name, had no clear connection with the river that had brought me here.

The Rose and Crown in Biddulph Moor is the pub closest to the birthplace of the Trent, but it was also shut. It struck me that the great explorers — John Hanning Speke, for instance — would not have been wasting time trying to find an open pub when there was a river to be explored. So I decided to leave my loved ones and strike across farmland towards the tracer line of blue which the Ordnance Survey map represented as the infant stream. I strode through a meadow of ungrazed, uncut grass that came up to my waist. Pollen rose in visible clouds, and I was soon sweating hard, sneezing explosively and rubbing my eyes, trying to comfort myself with the thought that explorers must expect some discomfort. I came to a wandering line of stubby alders and willows, along which a tiny, dark brook flowed purposefully in the direction of Stoke.

Actually, this journey, like all journeys, had started somewhere else a long time before. I grew up in a village in Berkshire and its river was the Loddon, which gave its name to a graceful flower of the water margins, the Loddon Lily. It flowed around the eastern side of the village in two branches that were united a mile or so downstream before running into the Thames at Wargrave.

One branch powered the mill, then raced from the mill pool under the road and past a large, rambling white house. Below the house there was a wooden footbridge, then a boathouse half-hidden in ivy, where the punt was kept. A little way further down was a slow, deepish pool known as the Bathing Place, with a shelter beside it where you could change into swimming gear. The other branch, known as the Second Stream, had been savagely dredged in some flood relief scheme and flowed dark and sluggishly along the far side of the meadow.

The property was owned by friends of my parents and we – my brothers and I – had the run of it. So the Loddon became my first river, and although it is many years since I wandered those banks, the memories are clear. The Welsh poet R. S. Thomas wrote of 'letting the stream/comb me, feeling it fresh in my veins'.[1] I was combed by the Loddon from an early age. We spent as much time as we could beside it, on it and in it. I caught fish from it. I floated down it, stretched face down out on the punt peering into its flickering being. I swam it, hauling myself along the beds of thick, rough emerald weed, poking my head into the hole in the chestnut roots where the chub and barbel swayed in the gloom.

I dreamed and day-dreamed about my river, and not always about a ragged piece of bread crust bouncing down between the weeds and being engulfed by a chub (though there was a good deal of that). It flowed through my conscious and subconscious. It still does, although it has since been joined by all the other rivers that have delighted me.

I like lakes and ponds, too, anywhere fish live and breed. But to me there is often a melancholy in still water, or at least a quiet prompting to reflect seriously on life's vicissitudes. It is the dynamism, the restlessness, the inconstancy of the river that touch me more deeply. I was also, without knowing it, intrigued by the way the river worked.

*   *   *

The appearance of that band of water – of an entity responding uniformly to the force of gravity – is deceptive.[2] The cross-section is like the contents of a packet of dried spaghetti viewed from the front, except that all the strands are moving at different speeds in subtly different directions, as if fighting to outdo each other.

Maximum velocity is just below the surface in the middle, the minimum along the banks and the river bed. Along comparatively straight stretches of river, two currents develop side by side in a spiralling motion, drawing water in from the fringes, revolving it around and down and back to the margins.

As the river narrows or shallows or both, flow accelerates. Where the course is strewn with boulders, the speed of the individual strand can surge and diminish every few centimetres. A rock breaking the surface creates behind it a standing wave whose crest builds up and collapses, over and over again. Within that wave, the current is upstream, which is why salmon choose these sanctuaries on their ascent of the rapids. Other boulders hide smooth summits just below the surface, creating patches of glassy calm where the water molecules come to a momentary halt. Where the river bed is entirely smooth and even – over fine mud or polished rock – there is a layer of this so-called laminar flow.

The threads of water follow their own paths, tangling with each other, rolling over each other, twisting and unravelling, dispersing, accelerating and decelerating, tracking diagonally, straightening, going back on themselves to form eddies. They compete in a state of eternal, chaotic instability, their patterns never repeated.

The surface hints at this state of agitation. It is disturbed by ceaseless balletics. Tiny whirlpools pirouette and collapse. Whorls shunt sideways, dodging and weaving. Patterns of marbling shiver across the reflected light of the sky. Waves rise and tumble. Back currents suck their way up the bank as if to ambush the main current, then lose their purpose in scummy stasis. Leaves and flotsam ride the middle line between converging currents.

Then there is the sound they make. Poets are attached to the fancy of the river singing songs, making melody – Shakespeare's 'sweet music with the enamell'd stones'. They don't sing, any more than they play, dance, chuckle or chatter. Nor – if we are being precise and literal – do they speak to us at all; nor can a river be a dark brown god. Nevertheless the complex nature of moving water does evoke deep responses within us, some of which were explored by the French philospher Gaston Bachelard, in his book *Water and Dreams*.[3] To Bachelard water is the spring of being, the essence of motherhood. It swells seeds, thereby ensuring the continuous birth, and concentrates images of purity and freshness. It carries us and soothes us, cleanses, cures and revives us; but also, being made of tears, suggests sadness and images of death.

For some, the wonder of the physical world is concentrated into mountains. For others it is the seas, the troposphere, potholes, deserts, flowers, bees, bogs, birds. For me it has always been rivers, and always will be; and when I no longer have the strength to wade them, or even sit beside them, they will still be with me, their life in me.

Leaving aside the metaphysics, rivers are also workers. They did not shape our planet in the first convulsions of creation, but they made it fit for our use. They broke down the rock to make the soil on which crops might be grown and animals fed. They fed the earth, they gave the power to turn the stones to grind our bread. Even now they provide the water we drink and wash with, and in many places they are the source of the energy that heats and lights homes.

We could live without mountains and we could go to the forests to obtain wood. But we could not do without rivers. They satisfied the thirst of the first humans, and drew the animals to them so there was food to eat. Rivers showed them how to find their way through the landscape, how to get home. They nurtured the first settlements, giving birth to the notion of society. Settlements grew into villages and villages into towns and towns into cities, always beside or near moving water. To build monuments to their gods, men turned to rivers. The bluestones of Stonehenge were quarried in Wales, shipped around the coast then floated up the Avon. Abbeys, priories, cathedrals, temples arose beside rivers because that was the only way the rock hewn from distant hillsides could be brought to the places deemed appropriate by the priests.

Invaders struck along rivers. Castles were built to defend river crossing points. When thoughts turned from making war to seeking trade, eyes focused on the river, too. Rivers separated realms, acting as lines to be defended or breached. But they also connected places and peoples, and the building of a bridge could symbolise the proffering of the hand of friendship, or at least a proposal of partnership.

They were the arteries and capillaries of the physical world. They also reached within, to touch the spirit, the divine. It was written in Genesis: 'And a river went out of Eden to water the garden; and from thence it was parted and became into four heads.'[4] These were the rivers of Paradise – Pison, Gihon, Hiddekel and Euphrates – made not just to water the garden of innocence but the whole of creation. For Walter Ralegh, locked up in the Tower of London with plenty of time to ponder such matters, the descending courses of the four rivers told the history of the world.[5] They measured the distance humankind had travelled, and the distance to be retraced if Eden were ever to be reclaimed.

To the philosophers, rivers seemed to fuse the human and the divine. Their element rose from within the ground and fell from heaven. They offered an analogy of divine purpose: infinite diversity sprung from a single design. It was obvious that, to succeed, man must placate and honour the river in order to gain access to its power and potential. Thus Aeneas sails into the Tiber in accordance with the prophecy of Helenus the seer that he will found a new city where he finds a white sow with thirty young.

The vision is fulfilled and Rome is born.[6]

The spiritual potency of the river is – or should be – at its purest and most holy at the point of its birth in the fountain or spring. There is the source of all that is exalted: love, goodness, beauty, justice, honour, tears. The idea is central to the Book of Revelation: 'And he showed me a pure river of water of life, clear as crystal, proceeding out of the throne of God and of the Lamb.'[7] John Stewart Collis, in his marvellously eloquent book *The Moving Waters*, wrote: 'Here is the pledge of purity; here is goodness made manifest; here is untarnished beauty; here we shall find truth.'[8]

That's how painters and poets would prefer it: the one source, the birth of the stream, life issuing forth from the bosom of Mother Earth. Unfortunately, the hydrological reality rarely corresponds to the aesthetic ideal. More often than not, there is no single point of emergence. The source is elusive, ambiguous, mobile. It takes some finding, and – as I was to discover with the Trent – tends to be something of a disappointment when finally located. The Canadian ecologist E. C. Pielou captured the downbeat truth perfectly: 'The majority of rivers begin at an indeterminate point in a slight depression in the ground where groundwater happens to ooze out as a gentle seep.'[9]

Chapter 2

Out of the Ground

The official source of the Trent, long ago

Before I could begin the journey in earnest, I had to be sure that I was on the right path. The stream I had found looked promising. From the Ordnance Survey map it appeared to be in about the right place and was certainly flowing in the right direction. But I had to find the source, whether it was a gushing spring or an ooze of groundwater. The river's course was a story which had to be followed from the very beginning. The narrative would take me downstream, but first I had to go up.

I tracked the stream around the edge of the meadow and up the slope towards a lane that ran south from Biddulph Moor. It became very small indeed, and disappeared into a clump of trees, on the other side of which was a fenced compound. As I forced my way through, a tremendous din of barking, howling and growling erupted. I glimpsed dogs leaping against the mesh of cages. I battled my way past the kennels and came to a hedge, with the lane on the other side. I established that there was no standing or moving water in the field beyond the lane, and therefore deduced that the dark and squelchy corner beside the dog compound must be the source.

I retraced my steps to where I had first met the stream and set off with it. It twisted this way and that between steep folds of rough pasture and I had to leap back and forth to stay in touch. While executing one of these leaps, I noticed with dismay that my stream was about to be joined by another, noticeably bigger. I had been deceived, not for the first time.

The road from Edinburgh south to Carlisle is kept company along the approach to Moffat by the upper reaches of the Tweed. The road climbs steadily, reaches a crest near a mighty scoop in the landscape known as the Devil's Beef Tub, then drops steeply into Moffat itself. A little way short of that crest is a lay-by where there is a tableau about the story of the Tweed and its course to Berwick away to the east. It states: 'In a field across the road, among pools and trickles in the wet ground, the Tweed starts its journey.'

John Brown, a doctor in these parts in Victorian times and the author of the popular *Rab and his Friends*, reported peering into the spring and seeing 'a gentle swelling like a hill of pure white sand . . . on which was visible a delicate column rising and falling in graceful measures as if governed by a music of its own'.[1] It is alleged that Coleridge looked into it before expressing the hope that:

> Nor ever cease
> Yon tiny cone of sand its soundless dance,
> Which at the bottom, like a Fairy's Page,
> As merry and no taller, dances still,
> Nor wrinkles the smooth surface of the Fount.[2]

Apparently Tennyson also visited the spot, and saw 'the spring, that down/From underneath a plume of lady-fern/Sang, and the sand danced at the bottom of it'.[3] But I can't help wondering if any of them actually took the trouble to squelch across the bog and heather to look for themselves, or whether they were fed the story in the old inn at Tweedshaws.

Whichever way, there is no crystal pool there now, or dancing sand. The hillside is one big, wet sponge of bog and moss, and the alleged source is one saturated patch, emerald with sphagnum moss. But if you cross the quaking meadow, beyond the reach of poetic curiosity, you can find flowing water, a hand's-breadth wide, gurgling between and beneath the clumps of bog grass. I followed it uphill until I reached a flattened cup of ground edged with stone and dark moss from which a finger of water descended. I stood beside it and as I looked down the valley, the nature of the deception became clear. The miniature stream that I was beside – known as the Cor – is clearly the primary source. Tweed's Well is for poets and the Tourist Board.

Away to the south-east, across an empty waste of heather and hill, Scotland ends and Northumberland begins, and here another fluvial puzzle awaited solution. The great river of north-east England is the Tyne. Its two branches join at Hexham, of which the northerly – coming down from Kielder Reservoir – is acknowledged as the senior. But there are complications at its top end. Reputable cartographers maintain that its true source is a rocky, acid little stream called the Deadwater Burn, which cuts down from high on Peel Fell, runs under the lonely road leading from the Scottish border to Kielder itself, and is joined by two lesser flows to make the Tyne.

But you will get a different story if you stop at Deadwater Farm and talk to the family who have managed this harsh land for two centuries and more, and therefore probably know a thing or two.

Mrs Hall, who was most emphatic on the subject, pointed me towards the corner of a field just off the road and just inside English territory. The light was fading and I was initially misled by a tiny black brook flowing through a block of conifers beyond Mrs Hall's field. I stumbled up and down this for some time until it dawned on me that it was going the wrong way altogether.

The next morning I walked upstream from where the Tyne – which is very small but indisputably the Tyne – passes Bellsburn-foot. The path went along a firm, grassy embankment that used to carry the old railway (closed in the 1950s) from Hexham north to its junction with the Carlisle–Edinburgh line at Riccarton. I crossed Deadwater Burn and looked up to the distant fell top whence it came. I had been reading Claudio Magris' rich and wonderful book *Danube*, in which he discusses the arguments that have raged for centuries over which of the three arms that unite to form the Danube at Passau should be regarded as primary.[4] To settle the matter, Magris whimsically invokes 'a branch of science, that of perceptology', defining its core principle thus: 'If two rivers mingle their waters, the one to be considered the main stream is the one which, at the point of confluence, forms the larger angle with the subsequent course. The eye perceives the continuity and unity of that river and perceives the other to be its tributary.'

On that basis, it was clear as daylight that Deadwater Burn must be relegated to the status of tributary. I pressed on past the platform of what used to be Deadwater Station, and came to a V formed by two minuscule burns. The one on the left soon became too feeble to produce any flow at all and lost itself in bog. But the other kept moving. I followed it across a saturated meadow. Several times I stopped, fearing it, too, had given up the ghost. But each time, as I stood still and listened, I could hear a watery tinkling; and there it would be, trickling around a clod of bog grass supported on black peat and over a miniature tumble of smooth, bronze pebbles. It went under the road, and it was as Mrs Hall had promised. Ahead of me was a hollow in the ground, and water was seeping out of it, downhill, heading for the sea.

In August 1794, Samuel Taylor Coleridge and his new best friend, Robert Southey, set off to walk from Bristol to Bridgwater. They were both young, energetic, and gripped by furious enthusiasm for the ideals of Pantisocracy, which envisaged a community of right-thinking men and women living a pastoral idyll far away from the shores of corrupt Britain, sharing everything, discussing anything and everything. They had practical details to settle, such as where the Rousseau-esque paradise was to be found, and who was going to pay for acquiring it. On the way they stopped for the night, and shared a bed, at the inn in Cheddar, and took time to inspect one of the few genuinely tremendous natural wonders in the land.

Two hundred years ago the wonder of the 450-foot-deep gash in the south-western flank of the Mendip Hills known as Cheddar Gorge was undefiled. The twisting road down into it was in perpetual shadow, squeezed between sheer cliffs of limestone honey-combed with black, dripping caves. From one burst a crystal stream, which raced beneath the road and lost itself in an enigmatic pool.

It did not take long for the tourist trade to ruin Cheddar Gorge. As early as 1912, Edward Hutton, the author of *Highways and Byways of Somerset*, was railing against 'the rabble of touts, the confusion of bristling advertisements and the air of Bank Holiday that have overwhelmed the town'. Matters have not improved since. Even at seven o'clock on a brilliant summer's morning, before the arrival of the first coaches, Cheddar had an irredeemably tacky air about it. Wretched, jerry-built concrete structures crowded around the entrances to the caves. A hoarding offered a competing attraction, Cheddar Man and the Cannibals, with the chance to execute 'your own prehistoric cave paintings, and ... make the shocking discovery: we were all cannibals once'.

The caverns and the emerging river dwelt in Coleridge's mind.

He became engrossed with the poetic possibilities suggested by moving water. He wrote in *Biographia Literaria*: 'I sought for a subject that should give equal room and freedom for description, incident and impassioned reflections on men, nature and society, yet supply in itself a natural connection to the parts, and unity to the whole. Such a subject I conceived myself to have found in a stream, traced from its source in the hills.'[5]

Coleridge's friendship with Southey, and the Pantisocratic ideals, collapsed in short order. But the impression left by the visit to Cheddar did not fade. Coleridge was searching for the continuation of the sacred stream, and in time he found it.

Mystery and magic sit heavily upon the minute church of Culborne and the wooded combe that hides it on the coastline between Porlock and Lynton. The church is no more than twelve yards long and four wide, with room on its austere pews for a congregation of thirty or so. The smell inside is of the sea, old wood, candlewax and hymn books tinged with damp. The rubble walls are twelfth-century, the two-light window in the north wall even older. Outside, a stream bubbles past the graves of the departed. The ancient woods of ash and oak press in from all sides, and the sea murmurs below. It's easy to understand why the monks who sailed here from Wales 1300 years ago looked no further.

Coleridge knew Culborne well. The walk between Porlock and Lynton was a favourite of his, and he had tramped it more than once with Wordsworth. But on this occasion he was alone – Coleridge himself said it was October 1797, although it may have been a year or two later. He was on his way back to his home at Nether Stowey when he fell ill and stopped at a farmhouse, almost certainly Ash Farm, which is above Culborne. He dosed himself with two grains of opium and – according to his own account – fell asleep while reading a passage from *His Pilgrimage*, an early seventeenth-century collection of travellers' tales by Samuel Purchas.[6] The passage began: 'In Xanadu did Cublai Can build a stately Pallace encompassing sixteene miles of plaine ground with a wall, whereine are fertile Meddows, pleasant Springs, delightful Streams . . .' In a 'kind of reverie' Coleridge began writing:

> In Xanadu did Kubla Khan
> A stately pleasure dome decree;
> Where Alph, the sacred river, ran
> Through caverns measureless to man
> Down to a sunless seas.

Forty-nine lines later he was famously disturbed by 'a person on business from Porlock'.

If you stand in the little lane that runs past the gate leading down to Ash Farm and look towards the sea, you can still experience something of the spell that gripped the poet's imagination. Smooth meadows slope down either side into the dark, wooded gulley enclosing the church and the stream. Beyond, the sea – colourless if not sunless – merges into the sky. Coleridge borrowed the oriental setting from Purchas, but his inspiration came at least as much from this magical, enchanted landscape, its 'deep, romantic chasm' pulsing with associations of fertility, its sacred river drawn from the black holes of Cheddar and the soggy slopes of Exmoor.

From the start men saw magic in rivers, and invested them with numinous qualities. When flowing in abundance, they gave abundance. In flood, they brought death and destruction. When they dried up, the crops and the animals and the people perished. Whatever could be done to keep them in a good mood – prayer, praise, libations, sacrifice – was done.[7]

In Babylon the blood of a ram was smeared on the walls of the temple to welcome the New Year, while its body was consigned to the Euphrates. The original Egyptian god, Hapi, was depicted as a male with hanging breasts from which the life-giving waters of the Nile gushed. Along the river's course were placed stone tables carved with runnels into which ceremonial wine was poured as an offering to Osiris. For the Egyptians, he and his sister/wife Isis were the providers of all life and all blessings, the chief of which being the mighty, fertilising river that came from an unknown place and made the desert habitable.

The Nile was acknowledged as the primary sacred river of the ancient world. The Greeks claimed it as a brother for their foremost river, the Achelous, 'the king of all the rivers'. They had nothing to match the Nile for scale; their rivers were much smaller, but just as crucial to the business of sustaining life. The answer was to give them all divine status so they could function both as geographical features and mythical entities. Thus Achelous the river rose (and still rises) in the mountains of north-western Greece and flowed out into the western end of the Gulf of Corinth, opposite the islands known as the Echinades. Achelous the god fought Heracles for the hand of Deianeira, but, despite assuming the forms of serpent, bull and finally half-man/half-ox, he lost and had his horn ripped off, which was then filled by the naiads with fruit and flowers and turned into the *cornucopia*, the Horn of Plenty.

Only the Alpheios could rival the Achelous in the aquatic pantheon. In its river role, it was used by Heracles to clean out the Augean stables. As a god, he was smitten by passion for the nymph Arethusa when she stripped off after a hot day's hunting and swam naked in one of his pools. He chased her across plains, mountains and chasms to Ortygia, where Artemis heard her cry for help and changed her into a fountain; whereupon Alpheios reverted to his watery form to be united with her.

At the end of the *Phaedo*, Plato's dialogue on the nature of death and the fate of the soul, there is an account of how all rivers, springs and seas have their origin in Tartarus, and are fed by Oceanus, which (and who) both encircles the earth and flows beneath it. Here, too, are the rivers of hell: Acheron, Pyriphlegethon,

Cocytus and Styx, in which humans are purged after death; and Lethe, the waters of oblivion. The Roman poets took Plato's system and grafted it on to their own cosmology. Thus, in the fourth of Virgil's *Georgics*, Aristaeus, heartbroken at losing his beloved bees, is taken by his mother, the sea nymph Cyrene, down into her watery realm: 'Now he gazed at all the great rivers flowing beneath the earth, separated in their places; Phasis and Lycus and the fountainhead whence deep Empeus breaks forth, whence Father Tiber and whence the Aniene's streams; and the golden Po with its bull's face and twin horns . . .'

Simon Schama has elegantly demonstrated in *Landscape and Memory* how the Romans also appropriated the Nile, absorbing its mythology into their own. But the Jews, understandably, turned against it. Led from captivity in Egypt, they came to the Jordan. 'Behold,' their god warns the Egyptians, 'therefore I am against thee and against thy rivers and I will make the land of Egypt utterly waste and desolate.'[8] This voice is like the Jordan itself: clear, powerful, purifying; quite unlike the muddy, meandering Nile.

The depth of the pools and the height of the waterfalls of the stream that flows down Culborne Combe through Coleridge's deep, romantic chasm are to be measured in inches. It is too small to have a name, and its history is short and simple. It seeps from a slope on Exmoor, finds its way down a convenient crease in the landscape and escapes into the sea. There are hundreds like it born on the bare, purple heights. Some go north, others take the longer way south. The moor looks solid enough but it is a giant sponge. It soaks up the rain, stores it in its mantle of peat and the network of caves, cracks, fissures and basins beneath; and issues it forth in whichever direction the vagaries of land formation dictate.

These wrinkles in the landscape can have important consequences. The newborn Trent's nearest neighbour – in the next cot, you might say – is called Bosley Brook. Between them is a ragged, toothy hump called Robin Hill. It is Robin Hill that sends Bosley Brook on its inconsequential way north to join the Dane at Congleton, and the Trent south on a 170-mile journey that will take it on a course shaped like a fish hook through the heart of Midland England to the North Sea.

All this I could get from the map, but the unresolved matter of the source nagged at me. I had come close that first day but not close enough, so I came back to look again. By then I had been with the river all the way to its end, but I could not think of the journey as complete until I had stood beside the birthplace.

I returned to the point where I first realised my mistake, then followed the senior flow upstream as it twisted between bushy alders and willows and past clumps of dark holly. Biddulph Moor reappeared on the ridge ahead, and the stream seemed to become unsure of itself, separating into threads of water that vanished into the grass and weeds.

Getting near

I traced one of these threads to a pipe poking out from a brick culvert running under the road through the village. The liquid dribbling from it was grey and smelled of drains. It spattered into a nettle-choked hole with crumpled drinks cans and a discarded packet of Golden Virginia lying around. Surely this smelly, sordid spot couldn't be the start of it? I looked around. A footpath of sorts led off to the right through nettles and brambles that reached up to my chin. There was a fence at the end with the legends 'Trent Head Well 130m' and 'Source of the River Trent' carved into it. A man working in his shed directed me back into the field I had just skirted. After wandering around it for a while, I almost fell into a brick-lined hole overhung by a cluster of hawthorn and overwhelmed by other vegetation. Hidden in the tangle was a black metal sign bearing the words 'Trent Head'.

Nearer

At last, I thought, thrusting aside the thistles and jumping down. There was an inch or two of water at the bottom, but the water did not move. There was no flow, in or out. Whatever anyone said, this could not be it.

One possibility remained: yet another minute watercourse emerging from a thicket of trees on the far side of the field. I traced it to another field with some sheep in it. In the near corner, shaded by the spreading branches of a lime tree, there was a liquid gleam. I tracked it back through the thicket, out the other side, and onward. It was joined by the tainted water from the pipe, then by the damp ditch that would have taken the water from the official source if there had been any.

By now the flow was a respectable twelve inches across, its progress interrupted by miniature riffles and pools. It turned left away from the village, cutting past a guard of river-loving trees: alder, weeping willow, sallow willow, past sprays of buttercup and colonies of ragged robin and the occasional swaying foxglove.

Little stream

I reached the meeting of the little waters. This was it. I thought of Speke and the Nile and Lake Victoria: 'I no longer felt any doubt that the lake at my feet gave birth to that interesting river, the source of which has been the subject of so much speculation.'[9] I rather doubted if my discovery would ever lead to invitations to lecture at the Royal Geographical Society, but there was a satisfaction in proving the map-makers, council heritage people and local historians wrong, mixed with a sense of anti-climax familiar to hydrographers. A chipped concrete pipe, a weed-choked hole in the ground, a muddy dip in a sheep field – there was little here of the pure and the sacred.

A mile or so below Biddulph Moor the insignificant stream made from a dozen less than insignificant trickles and dribbles in the moist Staffordshire earth passes underneath its first road bridge. I stood on it, holding the rail, looking down. The water was reasonably clear, but the bed, composed of mud and stones the colour of gravy, made it appear dark. It wore a sombre look, as if it had just heard the bad news that Stoke lay ahead.

Chapter 3

Childish Things

Never again would the river be so close to the way nature had made it as it was above and below that first bridge. It was left to itself. No one paid it any attention, no one bothered it as it found its way down its miniature valley. At first the way was open. Then the cleft steepened as it entered a dark, ancient wood. I followed its progress from above, stumbling around and over beeches, oaks and pines, collapsed and sagging like warriors fallen in battle. I caught glimpses of the water and, when I stopped to get my breath, I could hear it.

I came to a high stone wall, green with a moist mantle of moss, held up by rampant rhododendron, which must have once enclosed the grounds of some long-vanished country house. I hauled myself over it and battled through more abandoned shrubbery until I came to an intersection of footpaths, one of which crossed the Trent on a wooden footbridge. Above it the water ran smoothly and silently over angled slabs of stone. Further down, the path led to a mysterious pool, shut in by willows and reedbeds. Gas rose from its bottom in streams of bubbles to meet tendrils of vapour curled above the surface. Unseen fish circled and swirled. The air was very still and warm, heavy with moisture and the smell of mud and vegetation. A heron stood on one leg beside a sunken branch, hunched and still, waiting for something.

*   *   *

Aristotle credited the proposition that water was the animating principle of life to his shadowy predecessor, Thales of Miletus, one of the Seven Sages of the ancient world.[1] Thales may have been one of those who, according to Herodotus, warned King Croesus of Lydia about the uncertainties of the human lot and the jealous disposition of the gods. He was said to have calculated the height of the pyramids by measuring the shadows they cast, and to have forecast the solar eclipse of 585 BC as a result of inspecting the records in the observatory at Babylon.

Everything he wrote was lost, so all we have about him comes from what those who came later had heard. The single certainty about him is that he spent most of his life in a place whose fortunes were shaped by a river and depended on the sea; which is maybe why he pondered water's place in the scheme of things. Miletus overlooked the Gulf of Lade, on the west coast of what is now Turkey, south of Izmir and the island of Samos. On the other side of the gulf was the mouth of a river marked on the map of Turkey as the Büyük Menderes, but more famous by its Greek name, the Maeander.

These days the ruins of Miletus are ten kilometres inland, and what was the island of Lade in Thales' time is mainland. Then, in the seventh century BC, it was a dynamic sea port, trading all around the Mediterranean, exchanging olive oil and wool for timber, fish, iron and other goods, a centre of wealth and culture. But even then the river's habit of dumping enormous quantities of silt around its mouth was causing trouble. The island was expanding, the water was getting shallower, the riverbanks were edging forward.

The Maeander was a byword for muddiness and proverbial for the tortuousness of its course across the plain of Phrygia. Its god, Maeandus, was a favourite son of Oceanus, and a passage in Ovid's *Metamorphoses* celebrates the beauty of his daughter, Cyanea, who 'strolled beside Maeander's winding banks, her father's stream, that turns so often back upon its course'.[2] Homer and Hesiod were among those who made mention of its extreme windiness. In time, the name followed a leisurely route into the English language.

The mechanics of a river's meanderings are complex and are determined by the energy in moving water.[3] Some of this energy is converted into heat and some is stored as the river goes on its way. But much is expended on assaulting and undercutting banks, moving sediment and redirecting the current. Say you are looking downstream, and the angle of the current is into the left bank. It eats at the bank, dislodging silt and gravel, but at the same time it is deflected to the right. The more material it removes, the greater the angle, until the deflection is sufficient to take it across to the other bank, where the process is duplicated.

The wonder of the meander is its regularity. The wavelength is almost invariably between ten and fourteen times the width of the river, averaging eleven. The wavelengths are similar in shape but not symmetrical, the differences dictated by the composition of the bank. Resistance to erosion varies continuously, both in place and time, and is much reduced when the banks are sodden and the river is in flood. Rivers with weak, sandy banks become wide and shallow with long meanders. Those flowing through rock are narrower and deeper, the meanders tighter.

In the natural state, a river's course is always changing. Sometimes the pattern of the meander remains much the same, although the actual position of the loops shifts little by little downstream. In other cases, the concave curve is attacked by the current at its centre and the dislodged sediment is deposited at the nearest outward bulge. The meander becomes more and more exaggerated until, one day, a flood bursts across the neck of the loop to create a new channel. The water left behind, trapped in the discarded loop, is an oxbow lake.

When Coleridge made music with the words 'Five miles meandering with a mazy motion/Through wood and dale the sacred river ran', he was talking about the visible, horizontal pattern. But this is complemented by a less obvious vertical one. The clue is that the water is deepest and the current strongest next to the undercut bank, and shallows with slackening force towards the opposite curve. The current is deflected sideways, as already described, but also downward to create a spiralling motion. The contours of the vertical meander are determined by the nature of the river bed. Where it is softer, the scouring is deeper.

Left to its own devices, the river will run in a sequence of pools and riffles to go with its sideways twisting and turning. Left to its own devices, it is a marvel of fluctuating dynamics and mazy motions. But it is rarely left to its own devices.

The Greeks saw the meander pattern as a design feature, enhancing the gift of the river by extending the area of benefit and extending the impact of floods dispatched when the gods were angry. It was evident to them that rivers enjoyed a special status in the created world. The mountains and hills were immovable. The forests were fixed. The ground was solid. Even the seas and the lakes had their appointed places. But rivers were simultaneously permanent and forever changing. For those early 'ingenious pursuers' the river offered a pleasing symbol for their quest for knowledge.

In his *Naturales Questiones*, Lucius Annaeus Seneca conducted a systematic search of the physical world for signs to assist self-knowledge. 'Let us study the waters of the earth,' he wrote. 'When you have discovered the true state of rivers you will understand that you have no further questions.' Seneca accepted Thales' thesis, that water was the primary creative element. It followed that water acted as an image of human nature. 'Individual drops of water are mirrors of ourselves,' he wrote.[4]

Seneca's Stoic caution about firm answers was the antithesis to the certainties of revealed religion. His Jewish contemporary, Philo

Judaeus, of Alexandria saw in the natural order, not a mirror of us and our failings, but a clear window on to the benevolent design of God. He concentrated on the Paradise myth: Eden and its four rivers. For Philo, these represented the particular virtues – prudence, self-mastery, courage and justice – and their waters were the image of heaven on earth.[5]

Jews and Christians identified the rivers as the means devised by a Supreme Being to nourish and water His creation. Walter Ralegh, reflecting in the Tower on journeys spiritual and temporal, saw water as the primary agent of divine will and the river as an analogy of heaven itself.[6] His contemporary, the poet/soldier/ philosopher Sir Philip Sidney, drew a comparable parallel: 'In waters we have the head of them in the earth, and the spring boyling out of it and the stream which is made of them both and spreadeth itself out far from thence. It is but one self-same and continuall and inseparable essence, which hath neither foreness nor after-nesse, save merely in order and not in tyme.'[7]

By the time of Sidney and Ralegh, the view of God's intentions in creating the world was changing fast. Medieval thinkers, awed by the mysteries around them, tended to a comparatively modest assessment of humankind's place in the created order. But as scientific inquiry revealed more and more of the complexity of the world and its universe, so modesty gave way to hierarchical assumptions. It was obvious that God had made the world, but why had he lavished such astounding subtlety and ingenuity on it? Back came the answer: it was for us.

For Plato, Aristotle and the others, one of the marvels of the world was that the rivers kept running and yet the sea never overflowed. An answer of sorts is given in Ecclesiastes: 'All the rivers run into the sea; yet the sea is not full: unto the place from whence the rivers come, thither they return again.'[8] Here is a very early version – elegantly expressed but profoundly vague – of what we call the hydrological cycle: the process by which the water delivered to the oceans is returned to the land and thence back to the sea.[9]

Plato had suggested that saltwater was taken back from the sea into an enormous cavern deep beneath the earth's surface, where it was purged of its bitterness until ready to be discharged from the springs.[10] Aristotle was dubious, pointing out that if such a receptacle was to supply all flowing water, it would have to be bigger 'or at any rate not much smaller' than the earth itself. Having himself observed the condensation of vapour into water, Aristotle hypothesised that the same process was duplicated on a vast scale within the earth, cold air sucked in and converted.[11] But he also accepted a role for the hills and mountains as storers of rainfall, from which the water could 'ooze and trickle together in minute quantities but in many different places' – although this process did not, in his opinion, explain why water continued to flow after long periods without rain.

Christian theologians leaned towards Plato's version. The standard Christian account, included by Bishop Isodore of Seville in the *Etymologies* that he compiled in the seventh century AD,[12] is actually borrowed from Pliny, who had written of the earth 'opening her bosom and water penetrating her entire frame by means of a network of veins radiating within and without . . . the water bursting out even at the tops of mountain ridges, to which it is driven and squeezed out by the weight of the earth'.[13] But Bishop Isodore did allow for a proportion, volume unspecified, to be drawn up from the surface of the sea by the agency of sun, wind and cloud.

The role of rainfall foxed them all. Seneca, living in sun-baked Italy, was unusually assertive: 'As a diligent digger among my vines, I can affirm that no rain is ever so heavy as to wet the ground to a depth of more than ten feet . . . How then can rain, which merely dampens the surface, store up a supply sufficient for rivers?' Curiously enough, a few years before Seneca's birth a Roman engineer and army officer called Marcus Vitruvius Pollio had produced a manual entitled *De Architectura*, in which he stated that rainfall *was* the primary source of springs.[14] But no one paid any attention to him, and the speculations of Plato, Aristotle, Seneca and the rest continued to muddy the waters for a very long time. A favourite refinement of the theory was that the subterranean passages for distributing water matched the veins and arteries created by God to move blood around the human body. As late as 1664, a Jesuit scholar, Anathasius Kircher, stated as fact (in his mammoth *Mundus Subterraneus*) that the energy to shift the water was provided by the tides.

Knowing no Latin or Greek could sometimes be an advantage. A Frenchman, Bernard Palissy — better known as the pioneer of ceramics — declared proudly: 'I have had no other book than the sky and the earth.' In 1575, when he was sixty-five, Palissy gave a series of lectures in Paris on aspects of the natural world including fossils, the composition of rocks and clays, and the origin of springs. The material was collected into a book called *The Admirable Discourses* which made its way across the Channel to London. In the section on water Palissy stated: 'When I had long and closely examined the source of the springs and the place whence they could come, I finally understood that they could not come from or be produced by anything but rains.'

Seneca may have laboured in his own vineyard but, compared with Palissy, he was an inexpert observer of the natural order. Palissy, wandering the hills and valleys around his home town of Saintes in south-west France, was untroubled by the seeming paradox of rivers continuing to flow in the heat of summer when the last rain was a distant memory:

> The water falling on the hills which are full of cracks and
> crevices, descends continually and meeting with no obstacle
> until it reaches some cavity in the stone and rocks, it then

collects at the bottom of that cavity until, having found a channel or opening, it issues in springs, streams or rivers, according to the size of the opening; and as this spring cannot naturally ascend the hills, it descends into the valleys.[15]

The potter had seen through one of the illusions of our planet: its appearance of solidity. Beneath the firm footing are shafts, fissures, holes, caves, huge echoing spaces of emptiness; and, as well, a vast, connected network of minute spaces and cracks within the rocks themselves. Some species of rock – principally limestone and chalk – are so porous that they can hold enormous quantities of water and are known as aquifers. But all rocks are opened up to some degree over the ages by the action of drops and trickles of water. It seeps down, seeking the weak spots in the structure of the rock, where the sediment grains from which it was formed were never securely cemented together.

This process of seepage is leisurely. The water sinks gradually until it reaches a route to the open air, to form a spring. Again, Palissy saw how it worked: 'Rainwater that falls on places that slope towards rivers or springs do not get to them so very quickly . . . So springs are fed from the end of one winter to the next.'

It was to be the best part of a hundred years before Bernard Palissy's intuition was confirmed. Another Frenchman, Pierre Perrault, carried out a series of laborious experiments on the Seine which demonstrated that the volume of its discharge into the sea was more than accounted for by the rainfall on its watershed. Perrault's *De l'origine des fontaines* (1674) was translated into English (as *On the Origin of Springs*), and read by the members of the Royal Society in London. He had done the donkey work. All that remained was for someone to make sense of it, and Edmund Halley stepped forward.

Halley – who, in that age, stands beside Wren, Hooke and Newton for originality and intellectual elasticity – had already cracked some of the darkest secrets of astronomy, charted the course of his comet, described the actions of the trade winds and much else besides. He now grasped two of the crucial links in the hydrological cycle: the transfer of water between sea and land, and the mechanics of rainfall. In 1691 Halley presented a paper to the Royal Society in which he explained the process with a clarity that, in the context of its time, seems almost miraculous:

> Those vapours that are raised copiously in the sea and by the winds are carried over the low land to those ridges of mountains, are there compelled by the stream of the air to mount up with it to the tops of the mountains, where the water presently precipitates, gleeting down by the crannies of the stone; and part of the vapour entering into the caverns of the hills, the water thereof gathers . . . into the basins of stone it finds, which being once filled, all the overplus of water that comes thither runs over by the lowest place and breaking out by the sides of the hills forms single springs.[16]

At a stroke Halley had made the underground hydrological cycle – whether in Aristotle's version, or Plato's as expanded by Pliny – redundant. He also explained two aspects of the planet that had long perplexed Christian propagandists: why so much of it should have been allocated to useless and dangerous oceans, and why the part designated for human use should have been disfigured by unsightly and unproductive mountains. With a slight adjustment of the focal lenses, it became possible to interpret the hydrological cycle described by Halley as a divinely organised system for balancing the volume of water discharged into the sea and that evaporated by the action of the sun and the winds and sent back to land. There could be but one answer to the question posed with such rhetorical complacency by John Wesley:

Who gave it [the water] that just configuration of parts and exact degree of motion which makes it so fluent and yet so strong as to carry and waft away the most enormous burdens? Who has instructed the rivers to run in so many vast tracts in order to water them more plentifully? Then to disembogue themselves into the ocean so making it the common centre of commerce? And then to return through the earth and the air to their fountainhead in one perpetual circulation?[17]

A little way beyond the murky pool where the heron was hoping for a late lunch, the Trent passed meekly under a bridge into the open expanse of Knypersley Reservoir. I met an elderly couple who were taking their elderly terrier for a walk. We agreed that it was a lovely day and that everything looked lovely. The woman asked me where I was heading. She expressed mild surprise when I told her, but no great interest. Her husband busied himself with the dog.

The river lost its way in the reservoir. At the southern end there was an overflow with a sign beside it: DANGEROUS WEIR. But it was dry, the mud on it caked and cracked. The water that should have been the river's had been diverted into a concrete culvert to feed the Caldon Canal connecting Leek with Stoke. So the Trent had no choice but to start all over again the other side, supplied by a miserable seeping from the mossy dam wall. It set off listlessly through scrubby fields, looping back and forth as if looking for help.

I kept the tea-stained trickle company as it dodged between the tentacles of housing stretching north from Stoke. We reached Norton Green, and a pub called the Foaming Quart, which was shut. The path took me past a sturdy house painted cream and flanked by dark, sagging barns. It was the last farm before the countryside finally capitulated to the sprawl of the city. Beyond it was the Caldon Canal, its water scummy and ochreous. The Trent skulked under the canal, then headed off into a belt of impenetrable vegetation. Rather than try to hack my way along it, I took the smoothly gravelled canal towpath.

I passed a man fishing on the far bank, a can of lager raised to his mouth above his motionless rod. Further on a middle-aged couple were having tea beside their narrowboat, the woman knitting in the dappled shade, her husband intent on a book of hymns. I watched a barge negotiate its way through a lock with a couple of inches of water either side.

Somewhere over to the left the Trent struggled on. Then something unexpected and heartening happened to it. It vanished beneath two bridges, the first carrying the disused Stoke–Leek railway line, the second the A53 road. As it came out into the light it quickened. For the first time it showed waving strands of ranunculus, the essential weed of a vital watercourse. The current twisted between the beds of weed, hastened over gravel into a high-banked pool. Shoals of minnows darted among the discarded condoms, cans and footballs.

A little way downstream children were paddling and thrusting a net into likely places. Beyond them, in a quiet, overgrown pool with a lorry tyre in it, I spotted three or four chub finning midstream. I sat and watched the water, and listened to it. For the moment it had its voice back.

Chapter 4

The Magnificence of Cities

General View of Potteries.

SOMEWHERE THE SUN IS SHINING BUT NOT IN THE POTTERIES.

Stoke calls itself Stoke-on-Trent, but it has never taken the river to its heart or given it much of a welcome. Only now are the city worthies getting round to offering it a little respect. Strategy papers have been drawn up and partnership projects forged, a commitment announced to 'the rediscovery, protection and regeneration of the city's river corridors', involving the creation of a River Trent Path. But thus far action has lagged far behind the fine words.

Maybe one of the reasons for the shabby treatment is that, although Stoke calls itself a city, it is nothing of the sort; rather an agglomeration of scruffy industrial towns sharing little other than geographical proximity and the condition of decay. Stoke has no centre, no common identity. You wonder who it is trying to fool: itself, or others?

All in all, at no stage in its history would Stoke have qualified to illustrate the thesis developed by Giovanni Botero concerning the relationship between rivers and cities. Botero was a Jesuit propagandist who, towards the end of the sixteenth century, wrote a book translated into English as *A Treatise Concerning the Causes of the Greatness and Magnificence of Cities*. His perspective was commercial, its fundamental assumption that 'God created the waters . . . for a most redie means to conduct and bring goods from one countrie to another.'

Botero looked around him. He saw northern Europe pulling away, economically and politically, while his own country, Italy, remained fragmented by ancient rivalries. The answer came to him: 'Their rivers runne, not between the mountains . . . but many hundred miles through goodly and even plaines . . . and so, by this winding and turning, they helpe divers cities and provinces with water and victuals.' Most favoured of all was Flanders: 'The Meuse, the Schelde, the Mosella, Tevora, Ruer and Rhone . . . runne plesantly and gallantly forthright and overthwart the province and mightily enrich it by the commodities of navigation and trafique of infinit treasur, which certainly wants in Italy.'

Italy's rivers – with the exception of the Po – were short and unruly. The Tiber caused Rome endless trouble with flooding until the city's fathers confined it within walls of stone (now concrete) and pushed it out of sight. The Arno was even more volatile, as recently as 1966 providing Florence with a savage reminder of its capacity for havoc.

Giovanni Botero was no scientist, but this did not inhibit him from dreaming up a pseudo-scientific hocus-pocus about the rivers of the north having an additional, God-given advantage. The water itself was different. It had been designed to enhance flotation, so that the Seine, for instance – despite being a 'meane river' – could accommodate ships 'of such bulke and carrieth burdens so great that he that sees it will not believe it'. Botero's conclusion was emphatic: 'Those cities must be the fairest and richest that have the most store of navigable rivers.'

The Trent creeps into Stoke furtively from the north, and escapes to the south looking surprisingly fresh and sprightly, as if relieved that nothing worse than indifference had befallen it in between. On the map, its course is a slender finger of green pushed hesitantly through the urban sprawl. To the west are four of the Five Towns of Arnold Bennett's novels: Tunstall, Burslem, Hanley and Stoke itself; to the east the last of them, Longton, plus the one he missed out, Fenton. None of them has shown any great desire to claim it or look after it. With the exception of the disconnected sections where paths and cycleways have been installed, it has been left to fend for itself, wriggling along between unkempt banks, confined within concrete walls, thrust beneath roads and around the back of factories.

Smoke

There was a time when all the six Towns had a use for it. The Potteries were The Potteries then, and a perpetual pall of grey smoke pressed down on the pot-banks and kilns, the chimneys, the mines, the brickworks and the blast furnaces. It was a place, in Bennett's words, 'unique and indispensable', defined by its usefulness and its ugliness. It was smelly, grimy, smoky and coated in dust and soot, but the nation needed it and valued it. Because of the Potteries — Bennett again in the first chapter of *The Old Wives' Tale* — 'you may drink tea out of a teacup and toy with a chop on your plate'. And the river's lowly part was to remove the waste matter disgorged by the process and the teeming population.

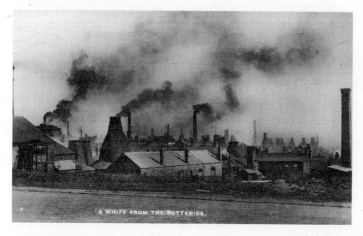

A WHIFF FROM THE POTTERIES.

Following the Trent through Stoke, you happen upon the left-overs: blackened pipes poking out from the base of crumbling factory walls, black, sticky ditches leading from abandoned yards and clay pits and heaps of spoil and slag. But that humble function has lapsed for good. As recently as fifty years ago, the river in Stoke was a toxic cocktail of human and industrial filth, biologically dead through the city and for many miles below. Now it is clean, its life paradoxically given back to it because no one needs it any more.

The Caldon Canal and the Trent keep close company for quite a distance through Stoke, and it was a lot easier to use the towpath

At work

than to hack my way along the river's twisting, tangled banks. The wreckage of the past was littered across the cityscape. Here and there an old brick oven rose from the rubble like a bloated teardrop, silent reproaches to those who had swept away the industrial landscape and omitted to put anything in its place. Swathes of old buildings had already been devoured by the bulldozers. Down quiet streets, warehouses and broken-windowed factories were awaiting their turn.

On the edge of Hanley the towpath took me around the boundary of a tree-fringed cricket ground. In front of the club-house lads were having catching and throwing practice. The green turf, the roped square, the white flannels, covers and sightscreens – all could have belonged to some tranquil village hidden in rural England. But on either side stretched desolate expanses of waste ground strewn with heaps of rubble, with the dark, greasy canal snaking between silent, abandoned wharves.

447. IT'S A PRETTY THING – STOKE-ON-TRENT !

A little way on, near the intersection between the A5009 and the A52, the two waterways parted company. The river crept south, the canal turned towards its terminus at Etruria, where the Wedgwood factory once stood.

Wedgwood the business is now situated a few miles down the Trent, at Barlaston, while the man himself has been appropriated by the heritage business. His good fortune was to be born into an age when one man could, if he had the energy and the big idea, make a place in his own image; and Wedgwood, more than anyone else, made the Potteries.

The area had long been known for its pots and jars and dishes and plates, baked in backyard charcoal-fired turf ovens. Wedgwood himself was born into a line of Burslem craftsmen celebrated for the quality of their earthenware. The brand identification was there, but there was one great obstacle to potential expansion. Everyone knew what it was, and everyone railed against it. Wedgwood did something about it.

JOSIAH WEDGWOOD F.R.S.
FROM A PAINTING BY SIR JOSHUA REYNOLDS P.R.A.

'I would advise all travellers,' Arthur Young wrote in 1770 in his *Six Months Tour Through the North of England*, 'to consider this country as sea and as soon think of driving into the ocean as venturing on to such detestable roads.' It was along these detestable roads that the packhorses had to slither and stagger bringing the salt, clay and other raw materials required by the Burslem potters, and taking away the crockery and vases and jars for which England was clamouring.

Wedgwood saw that what was needed was another form of transport: a canal, barges. At that time there was only one man in England for the task – James Brindley, known as the 'Schemer'.[1]

Brindley had been born in a bleak village in Derbyshire into a life of crushing poverty. Illiterate to the day he died, he surveyed his canals as he walked and planned them in his head as he lay on his bed. When Wedgwood went in search of him, he was already famous for having provided the Duke of Bridgwater with a canal to transport his coal from his mines at Worsley to the factories of Manchester and the port of Liverpool (it was said of the Duke that his conversation in later life rarely deviated from the subject of his canal and his coal, which, given that they were bringing in the stupendous income of £80,000 a year, was understandable, if tiresome for his companions).

Brindley came up with a scheme that would make the Potteries great. He proposed a navigation from Liverpool and Manchester via the salt-mining towns of Cheshire and the china towns of Staffordshire to reach the Trent above Nottingham, where it was big enough to take the barges. Canal and river would link the Mersey in the north-west to Hull in the east, and open the way to the heart of manufacturing England. With Wedgwood waving the flag, the potters needed little persuasion, the landowners, rival interests and Members of Parliament a good deal more. But in May 1766 the Act sanctioning the canal became law. Two months later Wedgwood himself thrust a spade into the earth to turn the first sod, Brindley wheeled it away in a barrow, and a sheep was roasted in Burslem marketplace amid scenes of public jubilation.

Work proceeded apace. The section south and east from Wedgwood's new factory complex in Hanley (which he called Etruria) stayed close to the Trent and presented no serious engineering problems. The northern link was slower and more troublesome. To reach Etruria – which it had to, since Wedgwood was the prime mover – the canal had to overcome the obstacle of Harecastle Hill, requiring the building of seventy locks and the digging of a one-and-a-half-mile tunnel through the top.

In the end Brindley did not live to see the whole route completed, but the benefits of the enterprise were apparent long before that. As each stretch opened, costs plummeted. The potters were soon paying thirteen and fourpence a ton for Cornish clay instead of thirty shillings, and were shifting their wares at a penny three farthings a ton instead of ten pence.

Brindley and Wedgwood changed the face of their corner of England, turning it from a medley of disconnected villages and small towns into an industrial powerhouse – or as John Wesley, whose brand of evangelical Christianity flourished there, put it, 'from a wilderness into a fruitful field'. Who cared that through the fruitful field flowed the ever more poisoned stream?

Even if I had had the time, which I hadn't, I had no great desire to visit Etruria. I had read Pevsner's *The Buildings of England*, which reported that everything bar Wedgwood's house and a single brick roundhouse had been torn down – the factory with its cupola and belltower, the streets and squares of cottages, the kilns, the warehouses, the whole lot. 'From his house,' Pevsner comments, 'he could look across landscape to the canal inspired by him and the works built by him. Now that view is all desolation.'

So I left the Caldon Canal to find its way through the Lord Nelson Industrial Estate to Etruria, and tracked the Trent as it threaded its way through the Station Industrial Estate and the Marcus Industrial Estate, under the traffic-choked A50 and through the Ravenside Retail Park into the leafy purlieus of Staffordshire University. A sign beside a footbridge announced that the area had been designated a nature reserve, and that a partnership had been established between the university and the civic authorities to promote the use of the river by otters and water voles. Standing on the bridge I counted in the river six tyres, three supermarket trolleys and one bath, but no mammals or fish.

I kept with it through the nature reserve, then lost it amid the roaring din of cars and trucks fighting their way through a maze of roundabouts and junctions bringing together the A52, the A5007 and the A500. It occurred to me that if I were an otter I would probably seek somewhere a little more peaceful. I crossed the Trent/Mersey Canal, dodged the traffic grinding along one of the A500's carriageways, and jumped over the safety rail on to the central reservation. Looking down, I spotted the Trent. It reappeared on the other side of the road, with a path beside it and a warning: YOU TAKE THIS PATH ENTIRELY AT YOUR OWN RISK.

Suddenly this was sylvan Stoke, or the nearest it got. The river, shallow, gravelly and thick with weed, swung to and fro through a meadow, shaded along the far bank by stately willows, leaning alders and the occasional towering copper beech. I came upon a gang of youths and girls, and was reminded of the sign. But they showed no interest in me, and I walked on. At the end of the path the river disappeared beneath a roundabout and flyover uniting the A500 and the A34. I was unable to tell if there was a path through the tunnel, and anyway teenagers with bicycles were gathered in the gloom in an ominous way, so I scrambled up and over more concrete and tarmac and took the A34 south.

By now it was after eight o'clock. The sun was sinking over the M6 motorway to the west, but was still pretty fierce. My feet were sore, my legs ached, my throat was parched and I felt drained of strength. I needed a pub urgently. One appeared on my left called the Staffordshire Knot.

Inside, a woman of mature years, with dyed white-gold hair and squeezed into a tight top and tighter skirt, was propping up the bar, lager in one hand, fag in the other. She watched me as I came up and ordered a pint. 'Watch out, it's fookin' Tiger Woods,' she said, cackling hilariously. Tiger Woods? I was flummoxed for a moment. Then I remembered my straw hat. It must be the hat. I nursed my cold, fizzy, nasty beer resentfully in a corner. Was this the proverbial friendliness of the Potteries, I wondered? Then I thought: she doesn't like me because I'm from somewhere else.

Chapter 5

North of South

The distinguished English geographer W. G. Hoskins claimed that a remarkable transformation might be experienced as the north-bound steam train (this was the 1940s) crossed the Trent: 'The landscape changes suddenly now: stone walls blackened by indus-trial dirt replace the hawthorn hedges of the past two hours, the air in the carriage is perceptibly colder and fresher, voices on station platforms are harder and speak in strange tongues . . . the traveller stirs from his torpor. He has crossed the Great Divide between the South and the North.'[1]

'The North' is always northern England, never Scotland, which is somewhere else altogether. The partition Hoskins refers to is embedded deep in the national consciousness and goes back a long, long way. As early as the late Iron Age coins were in circulation south and east of a line stretching from Dorset to Lincoln, but not beyond it.[2] The Romans separated their most northerly dominion into Britannia Superior and Inferior. Thirteenth-century official-dom was satisfied to regard the King's writ as applying *citra Trentam* but not really *ultra*. The distinction reflected economic and demo-graphic realities; thus Domesday recorded that virtually all English territory with a population of two and a half or fewer to the square mile was north of the Trent. In Chaucer's *The Reeve's Tale*, John and Allan come from 'fer in the north, I kan nat telle where' – in other words, enough said, they're from up there.

The Trent has long done service as the southern boundary of

that imprecisely defined area. In Shakespeare's *Henry IV*, the leaders of the rebel factions – Percy, Edmund Mortimer and Owen Glendower – meet at Bangor to work out how they will divide the kingdom once they have disposed of the usurper Bolingbroke. Smooth-tongued Mortimer lays the deal on the table:

> England, from Trent and Severn hitherto,
> By south and east, is to my part assign'd;
> All westward, Wales beyond the Severn shore,
> And all that fertile land within that bound,
> To Owen Glendower; and, dear coz, to you
> The remnant northward lying off from Trent.

But Percy, alias Hotspur, can read a map even if he does hail from Northumberland. He knows that the river's retreat north-east at Burton and north at Newark leaves the rich lands of Leicestershire and most of Lincolnshire to 'dear coz' Mortimer:

> Methinks my moiety, north from Burton here,
> In quantity equals not one of yours;
> See how this river comes me cranking in,
> And cuts me from the best of all my land;
> A huge half-moon, a monstrous cantle out.
> I'll have the current in this place damm'd up,
> And here the smug and silver Trent shall run
> In a new channel, fair and evenly:
> It shall not wind with such a deep indent
> To rob me of so rich a bottom here.[3]

It would not have occurred to Hotspur that coming from the other side of the smug and silver Trent made him culturally inferior to Mortimer or any other southern noble; and you suspect that had anyone made such a suggestion he would have challenged it at the point of his sword. Hotspur's objection to the proposed division was not that he was getting the substandard part of the kingdom, but that he was being cheated on acreage.

The industrialisation and urbanisation of England from the end of the eighteenth century effected an imprecise partitioning. Writers of fiction began to notice that the towns and cities of the north seemed to be much smokier, dirtier, noisier and fuller of people than those in the south (excepting squalid London).[4] Disraeli toured the industrial north-west and wrote a novel, *Sybil*, to which he gave the subtitle *The Two Nations*, thereby acknowledging the emergence of an urban underclass spawned by the factory system. In *Hard Times*, Dickens pictured the industrial town of Coketown (based on Preston), its streets straight and hopeless, its smoking chimneys, the clouds overhead heavy with soot, 'a black canal and a river that ran purple with its smelling dye'.

The image of the north as being different began to acquire extra layers of stereotyping. Away from the grimy towns, the country-side was harsher and bleaker, the moors and bare fells far removed from the gentle downland of Wiltshire and Berkshire. The climate was more severe, and the alliance of weather and geography made for a different kind of people: tougher, harder, grittier, more down-to-earth, less given to sentiment. The north could be an unsettling place for a southerner to find himself. In *Wuthering Heights*, the force that Heathcliff draws from his savage moorland environment enables him to overwhelm the refined and sensitive Linton. Northerners also seemed to display, at least to each other, a warmth and openness unknown in the leafy Home Counties, even if – as George Orwell alleged – they had become so accus-tomed to ugliness that they no longer noticed it.

Orwell, who was born in Henley-on-Thames, Oxfordshire, crossed the great divide in the 1930s en route for Wigan. 'When you go to the industrial North,' he recorded solemnly, 'you are entering a strange country. This is partly because of certain real differences that do exist, but still more because of the north-south antithesis which has been rubbed into us for such a long time past.' Having noted it, Orwell sets about some vigorous rubbing of his own. He claims to have identified 'a curious cult of Northernness, a sort of Northern snobbishness . . . A Yorkshireman in the South will always take care to let you know he regards you as an inferior . . . he will explain that it is only in the North that life is "real" life, that the industrial work done in the North is "real" work, that the North is inhabited by "real" people.'[5]

Less than twenty years ago, one of the architects of Thatcherism, the Trade Secretary Lord Young, offered this historical insight: 'Until 70 years ago the North was always the richest part of the country. I try to encourage people to go north, that is where all the great country houses are because that's where the wealth was. Now some of it is in the south. It's our turn, that's all.'[6]

By the time Young uttered those brutal words, the great southward shift of economic power was almost complete. Now, of course, the mines, the blast furnaces, the shipyards, the steelworks – the outward and visible signs of northern industrial might – have pretty much gone. The slagheaps observed by W. G. Hoskins have been levelled or landscaped into grassy knolls, the stone walls have been cleansed of their grime. The strange tongues endure, but are they really stranger than those of Somerset, Cambridgeshire, east London or even Essex? And can the air ever really have become colder on one side of the river than on the other?

Even in Hoskins' day, the notion of northernness was sustained by much unthinking myth-making. The clichés abounded: cloth caps, ferrets, whippets, pigeon-lofts, small men packed shoulder to shoulder on rain-swept football terraces and so on. These days northernness seems more a state of mind than a reflection of a shared identity or fate. As cities like Manchester, Leeds and Newcastle reinvent themselves in the contemporary, global mould, the physical evidence of a north–south divide is disappearing fast.

For how much longer will the traveller crossing the Trent by train, or entering a pub on the outskirts of Stoke in search of a healing pint, feel that he is an interloper in a strange land?

For the time being, though, it is still a little different. Bed and breakfast that night cost me £20, a rate unheard of in my part of the world since Lord Young's day. The place was another twenty minutes' trudge from the Staffordshire Knot and called itself a 'hotel'. But the friendly and welcoming lady who ran it, a refugee from Birmingham, dismayed me with the information that she provided neither food nor bar. So, having dumped my rucksack, I hauled myself across the main road, through the nondescript settlement of Hanford to a huge roadside hostelry full of families looking at each other over enormous plates of chips.

I got an enormous plate myself, filled with chips, steak, baked beans, mushrooms and two fried eggs, all for £9.95. I felt far from home, but the beer, Bass from downriver at Burton, was delicious and I was glad that I was almost done with the walking. I needed to be on the river, or in it. Being beside it wasn't enough. I wasn't part of it. The next day I would, with luck, be afloat.

Chapter 6

Holding Their Noses

In 1859 a Nottinghamshire artist called Hilton Lark Pratt executed a painting of Trentham Park, the estate – one of them – of the Duke of Sutherland just south of Stoke. His vantage point was close to an obelisk which rises from the woods at the southern end of Trentham's lake and supports a bronze statue of the first Duke. There is an inscription on the obelisk honouring the Duke as 'an upright and patriotic nobleman, a judicious, kind and liberal landlord . . . a public yet unostentatious benefactor who opened his hand wide to the distress of the widow, the sick and the traveller', a panegyric which might well have been disputed in the Highlands of Scotland, where he is chiefly remembered as the instigator of the infamous Clearances.

The focus of Pratt's picture is Trentham Hall, the vast mansion commissioned by the 2nd Duke of Sutherland, designed by

Trentham as was

53

Charles Barry, and completed a few years earlier at the colossal cost of £250,000.[1] The building stands gleaming beyond the fountains, statues, gravelled walks and sumptuous borders of the formal gardens. In the foreground is the lake originally created by Capability Brown, to the side the silver band of the Trent. All is nestled in pasture and woodland, and at first sight everything seems perfect. But the young Pratt did not allow himself to be carried away by the splendour of the spectacle. In the far distance he darkened the sky with smoke and sketched the silhouettes of factory chimneys. The grime and soot of the Potteries were closing in.

Twenty years before, the Trent had flowed directly into the lake. But the ever-increasing load of sewage and factory muck it carried with it made the area smell most evilly, to the distress of the Sutherlands and their aristocratic guests. They had it diverted to the east, but the respite was short-lived. A smallish stream was being required to remove the combined human and industrial waste of six towns and its protest was to pollute everything it touched, and to stink to such a degree that people of high birth could not stand it. In the words of the local newspaper: 'The pools of the princely grounds of Trentham are becoming the cesspools of the Potteries.'

Toxic Trent a century ago

The lustre of the great house and its princely grounds faded fast. The Sutherlands increasingly shunned it in favour of one or other of their estates in Yorkshire or Shropshire, their mansion in London, or their castle in north-east Scotland, Dunrobin. The 4th Duke, hard-pressed by death duties and the cost of a scandalous settlement with his father's mistress, resolved to get rid of it. The contents of the library were sold in 1906, and the next year the paintings – the Holbeins, Hilliards, Lelys and the rest – went the same way. His Duchess – popularly known as Meddling Millie because of her energetic commitment to good works and her campaign against the pottery industry's use of toxic lead in the paints and glazes – complained: 'Strath has a mad lust for destruction on the plea of death duties ... Strath is a monomaniac ... a pitiable figure, mooning about like Scrooge and muttering about money.'[2]

He offered Trentham to Staffordshire County Council as a training centre for teachers, but scuppered the hopes of any deal by insisting that women should not be admitted. The remaining contents were sent off to Christie's, and in the late summer of 1911 the destruction of the great house began.

A few vestiges of Barry's vision did escape the sledgehammers, and stand today looking rather pathetic and forlorn. A section of the south front rises from rubble and rusty railings, with the top of the stableyard clocktower appearing above its flaking stonework and crumbling balustrade; and the west arch also survives, laden with armorial bearings. The new and exceedingly gloomy church that Barry provided for the Sutherlands is still there, and in the gardens Perseus continues to brandish the head of Medusa. The gardens have been restored, and across the stream buildings more attuned to contemporary tastes have arisen: a shining garden centre, rows of timber shops and cafés, a black smear of tarmac beyond for the cars and coaches.

Clocktower

For a brief half-century Trentham had been the grandest new country house in England. Disraeli presented a thinly fictionalised version of it in his novel *Lothair*: 'It was an Italian palace of freestone; vast, ornate, and in scrupulous condition, its spacious and graceful chambers filled with treasures of art, and rising from statued and stately terraces.' The grandeur that so dazzled him has been erased, and no more than a dim aftertaste can be obtained from the pictures, prints, Barry's sumptuous designs, and the remnants.

But also banished – more recently – is the curse that had the Duchess and her fragrant friends reaching for their perfumed handkerchiefs every time the breeze came from the direction of the water. The lake is healthy enough to sustain large carp, and pike and tench. And the Trent itself runs clean, a little river again instead of an open sewer, which may be some consolation for the loss of the great house, and the chimneys and kilns of the Five Towns.

My friendly Brummie landlady at the B & B up the road told me that, a week or so before, she had been walking along the river and had come upon an otter munching a fish. She then produced a photograph of her husband holding an astonishingly large trout which she said he'd caught on a worm in a pool not far away. 'It's a rainbow trout, I think,' she said proudly. It wasn't. It was a brown trout, at least four pounds, a wild fish. In the Duchess's day – indeed, as recently as twenty years ago – you would have been as likely to see a giraffe drinking at Capability Brown's lake as encounter otters and trout on the Trent.

After breakfast I pushed my way down through a head-high mass of grass and wild flowers to inspect the pool where the trout had been caught. I tried to follow the Trent downstream, but the way along the bank was blocked by an intimidating fence with a sign on it advertising the presence of guard dogs, so I went back to the main road and walked down it to Trentham, where my boat was waiting for me, together with my wife – who had taken the girls to stay with her parents in neighbouring Cheshire – and my father-in-law.

I had thought long and hard about the craft for my adventure. She had to be light enough for me to manage on my own, and – at a pinch – to be transported for short distances overland. She had to be big enough to take me and my supplies, and for me to sleep in should the necessity arise. At the same time she had to have the minimum possible draught so that she could be manoeuvred in shallow water. Most important of all, she had to be right for the river, aesthetically and philosophically. If a river could be pleased with a boat, this river had to be pleased with my boat.

Those were the requirements. I took them to my friend Jon Beer, who – as well as being an exceedingly accomplished angler and writer – is enviably talented with his hands. I knew that he had designed and made a punt for skimming up and down the minuscule River Cherwell, which runs past his home. So I asked him to make one for me, for the Trent, and he did: a punt, but no ordinary punt. She was fifteen and a half feet and a little more in length, three foot three inches in the beam, square-ended and flattish-bottomed but slender and sleek, made from sheets of plywood, with oak blocks at either end and a central mainframe of oak to hold her together. She was – she is – tough, durable, manoeuvrable and marvellously elegant: a boat for one man. One man and his boat.

I had painted her a soothing olive green, and as I walked down the A34 to renew acquaintance with her, I decided that she should be known as the *Trent Otter*. This was partly because the creatures were obviously about; but more to honour a long-departed angler called J. W. Martin, who used the pseudonym 'Trent Otter' for a series of books with titles like *My Fishing Days and Fishing Ways* and *Days Among the Pike and Perch*. He had long been a favourite of mine. I loved his honest, modest tales of great bags of chub and barbel and bream, taken in distant days on outlandish baits like bull's pith and boiled scratchings.

Under the gaze of Barry's clocktower, my father-in-law and I slid the *Otter* easily over a grassy bank into a foot or so of water. The Trent looked dark and rather sombre on account of its bottom of brown Staffordshire mud and stones, and it flowed lethargically between beds of thick, scummy weed. I loaded my gear and provisions into the front, then placed a plank athwart the stern to act as my seat. The launching was a low-key affair: a handshake from my father-in-law, embraces from my wife mixed with words of caution and admonitions to wear my life jacket at all times, as if I was embarking down the Zambezi or the Orinoco. I sat down and grasped my paddle.

I was glad it was a paddle. The original idea, since she was a punt, was to pole her. Jon Beer was very keen on this method of propulsion and, to show me how easy it was, we had a trial run from his cottage up and down the Cherwell. In his hands, the pole was an instrument of precision, and we flashed between the rushy banks, turning this way and that with a thrust here and a touch there. But when I tried it we careered from side to side, from left bank to right bank and back, and got nowhere. Jon smiled and said it would come with practice. I sweated and cursed and stumbled and got cross.

In the end I tried a paddle, and we did better, well enough to get back to his cottage. There he took the pole to demonstrate how pleasingly and usefully easy it was to execute a 180-degree turn with one thrust. As his end spun around, it struck the edge of a small landing stage and came to a halt. He went on and struck another part of the landing stage. He lay with one foot still in the boat, his torso splayed across the planks, gasping with pain. I went off the idea of punting.

Now, afloat at last, I applied the paddle. The *Otter* and I eased forward. Two more strokes and we ran aground on a gravel bank. I hopped out, pushed off, hopped back in. My wife and father-in-law waved encouragingly from the footbridge leading from the garden centre to Trentham Gardens, and I waved back. There is a photograph in which I look pretty much at ease, feet crossed, paddle across the knees, smiling with a confidence that I was far from feeling. They left, and I was on my own.

Looking happy

I felt the aloneness at once. Being on a river is very different from being beside it. It takes hold of you, and in that acceptance you leave behind your old, secure relationship with the solid earth. You no longer belong there. You pass it, it passes you, and the world is suddenly a different place.

Below the footbridge I soon disappeared from the view of the earthbound, into a dark, winding alley between trees. I could hear the earthlings as they strolled along the path between the river and the lake. Occasionally I caught sight of children running or couples guiding pushchairs. But I don't think anyone noticed me as I crept downstream, as often in the water coaxing the *Otter* off the bottom as in her paddling. The sky was blue through the leaf cover, but the sunlight hardly ever reached me.

I came to a ruined two-arched bridge. The channel above it was deep and lined with crumbling brick walls draped in ivy and mottled with moss, which suggested there had once been a mill here. The left arch, looking downstream, was intact but entirely blocked by a fallen tree and a raft of rubbish that had collected in its branches. The current was thus diverted down the right side over a smooth chute no more than four inches deep. I was pretty sure it wouldn't take me in the boat, so I jumped out and held the *Otter* at the stern as she was pulled over the chute into a swirling pool beyond. The surface was slick with greasy algae and my feet went from under me as I followed her. For a second I almost lost my hold, before I was able to heave her into an eddy at the side, get my breath back and settle my jangling nerves.

Leaving the lake and the statue of the Duke of Sutherland behind, the river emerged from its cloak of trees into green dairy land, and came to a standstill. It was hot and getting hotter, and paddling was hard work. I had fitted the *Otter* with rowlocks, so I took to the oars around a long series of bends, sometimes pulling in the conventional way, sometimes backing in order to get past the bushes, trailing branches, tree stumps and other obstacles. To the west I could hear the noise of the traffic pounding along the A34. Somewhere on the other side was Barlaston, where Wedgwood was now located, but I could see nothing of that either.

I stopped for lunch in the shade of an oak tree. The *Otter* swayed gently in the current and the river whispered by. It had taken me two and a half hours to get here, and the cars I could hear were no more than five minutes from Trentham. Two worlds, side by side – straight road, fast car; winding river, slow boat. It was as well I wasn't in a hurry. On the far bank was a golf course. Occasionally sweaty red faces would appear over creaking trolleys, and I would catch a snatch of conversation about the heat or the lie of a ball or whether to take a nine or a wedge. Then there would be a swish of club and the smack of contact.

Below Tittensor the river shallowed and quickened. Thick beds of water crowfoot hugged the gravel, breaking the flow into braided channels. I steered with the oars, facing front, and lost count of the number of times I had to get out to push us off. We went under the A34 again, and then for a third time. Below this bridge

Below Trentham

a swan which I assumed to have been battered and maimed by a rival lay against the bank, half in and half out of the water. The plumage along its body and neck was flattened and streaked with mud, and I could hear its gasping breath. Normally I don't care much about swans, but I felt sorry for it, and wondered if I shouldn't do something to end its suffering, like strangle it or hit it over the head with my anchor stone. But by the time I'd finished wondering we were past it, so I had to leave it to die slowly.

The town of Stone was over to my left somewhere. Stone makes quite a fuss of Brindley's Trent and Mersey Canal, but hardly spares a glance for its river, which is left to follow its inclinations through peaceful water meadows. So I declined to notice Stone, and drifted on, absorbed in the business of keeping the *Otter* on course. Then I had a shock. I met another human being coming towards me.

He had a beard and a safety helmet and a life jacket and a double paddle three times as long as mine. The rest of him was invisible inside his kayak. For myself, I would have been quite happy to have waved and exchanged a couple of watery pleasantries and passed on. But he was keen to chat.

He was, he told me, a member of the Stafford and Stone Canoe

Club. Canoeing – 'paddling', he called it – was evidently important to him, but he'd been out of it for a while, something to do with rib cartilages popping out where they shouldn't. He twisted and stretched and grimaced as he talked about the history of the club. They'd had a world champion once, he said, pausing for me to acknowledge what an extraordinary thing this was. He told me about Pauline, who'd been a legend. Pauline's idea of combining duty with pleasure was to paddle up the canal from Sandon to Stone, then paddle down the river back to Sandon to attend a committee meeting at the Dog and Doublet.

'She doesn't paddle any more, Pauline doesn't,' my new acquaintance said pensively. He cheered himself up with a couple of sudden sprints, then slowed down to warn me about the Trent Trots, a species of Midland dysentery that would certainly afflict me if I fell into the river in Nottingham. 'Drink at least two litres of Coca-Cola – that's the only treatment,' he advised. He began to tell me about the Japanese knotweed spreading along the banks. 'Of course at the moment the plants are all female, which means they can only spread rhizomatously,' he confided. 'But if we get the males as well . . .' He blew out his cheeks, suggesting that the scale of the catastrophe would be beyond mere words.

The headquarters of the Stafford and Stone Canoe Club was just below the bridge in Stone, next to a slalom course created by installing concrete groynes along the banks. Various colleagues of the bearded paddler watched as I manoeuvred my way past these obstacles, taking a few thumps and bumps on the way. One asked me where I was heading. The sea, I said. He shouted with laughter: 'Fantastic. Imagine you're on a time trial.'

By now I was hot, seat-sore, sun-blasted, thirsty, weary, and getting anxious about where and how I would spend the night. My plan, inasmuch as I had one, was to find somewhere to camp, then camp. But I am not naturally the outdoors type, and my experience of camping was slight, hence the anxiety.

I was lucky. The river flowed along the edge of a small village called Burston, where there was a steel footbridge over the river. Just above the footbridge was a big willow, with a miniature bay and a shelving gravel beach at its foot, just right for the *Otter*. I pulled her up, tied her to the tree and took a path up into the village. A little way on, a hump-backed stone bridge took me over the ochre waters of the canal. A couple were sitting in their garden on the other side, having a drink and watching the sun lower itself in the sky. The man went into the house and brought me a can of cold lager which I emptied in one long, ecstatic draught. They said they thought the field belonged to the big house further on. I went to the big house. They were having a barbecue in the garden and I could smell grilling meat. The lady of the big house said I was welcome to camp.

Millpond in Burston

Never having attempted the operation for real before, I felt I needed more beer first. I walked into Burston past a row of rustic cottages and a pretty millpond with an island in it. I wondered irritably why the pub couldn't be in the village rather than at its far end, but the irritation soon disappeared when I got to the

Greyhound. Four pints later I returned the way I had come. An elderly man emerged from one of the rustic cottages carrying a fishing rod and stool and we exchanged a few words about the contemplative man's recreation. He was hoping for a tench or two before his wife called him in; maybe one of the big carp if he was lucky.

It took me an hour and a half to get my tent up and secured to the ground. By then darkness was gathering. Sticky with sweat and suncream, and annoyed and ashamed at my ham-fistedness, I sank into my collapsible chair to cool and calm down. I watched the river as it curled around a high sculpted bend into my little beach. The bank was pitted with the holes of sand martin nests. The surface was molten silver, streaked with gold, whorled and marbled by the competing threads of water. In the course of that day the Trent had left its infancy behind. It had acquired depth, strength, purpose. It was aspiring to be a river.

I wriggled into my sleeping bag and thrashed around for a time seeking a position of comfort on the inflatable mattress. I took off my glasses and put them down, and immediately started worrying about rolling on to them in the night. I thrust them into the furthest corner of the tent, and lay listening. I could hear cattle tearing and chomping at the grass in the field beyond the far bank, and the occasional wet splatter of one of them crapping. Between them and me the river kept up a soft, liquid murmur.

Chapter 7

Recreating the Spirits

Sylvan Staffs

Over breakfast – a jam sandwich, hot, strong coffee – I watched the water again. The dew was heavy on the grass and beaded the outside of the tent with droplets. The swallows were already out, darting and swooping after the first hatching insects of the day. The sky was pale blue, promising heat again. The softest of breezes was nudging from the south-west.

Behind me the village was as silent as a graveyard. But the cattle were up and about, mouths tearing at the grass, tails twitching. Between them and me curls of vapour unfurled across the water as it pushed past the steep banks and hurried beneath the

66

footbridge. It never rested, urged on by the imperative defined for it by gravitational force: to reach the sea.

Looking at any river, it's not hard to understand why they should always have had special status. Long ago rivers were worshipped, feared, and watched with close attention. Some even wondered if the river might have been endowed with a basic watery intelligence. A sixteenth-century French humanist, Pierre de La Primaudaye, thought it behaved 'as if it had some sense and understanding, and that God had caused it to hear his voice and commanded it, as he commanded Man, to obey his ordinance'.[1]

We see rain falling and rivers flowing, but – as Bernard Palissy worked out – there are enormous volumes of water on the move out of sight. Every river is accompanied in places by hidden bands of water, usually where a previous meander has laid down porous sediments, riddled with minute channels through which water is able to percolate. Its rate of progress is rarely more than a few centimetres an hour. But flow it does, and these so-called hyporheic zones play their part in the river eco-system, feeding the roots of bankside trees and plants, and offering a home to infant stoneflies and various species of minute crustacea.

Gravity is the motor of all this restless dynamism. It gave the rivers the power to carve the valleys and divide the mountains, to break down the rock into soil to be laid down as land for cultivation, and to deposit it in the sea so that the habitable world was extended. This process, fuelled by rainfall, driven by gravity, has shaped the temperate parts of our world and kept them going. The price we have paid is the rivers occasionally getting out of hand.

This two-facedness did not bother the ancients. They took it for granted that the river – Homer's 'giver of good things' – would also be used from time to time as an agent of punishment. Pierre de La Primaudaye noted that 'it pleaseth God that they should overflow to chastise men by deluges and floods'. The belief in divine punishment administered in the form of natural disasters is less fashionable today than it once was. At the same time, we have tended to forget that living with a river and forcing it to meet our needs carry an element of risk. Every river on earth has a history of causing calamity. But only in our age have we become complacent enough to think that we have eliminated the risk. We build houses in floodplains and are astonished when they are flooded. A great city, New Orleans, is built below the level of the water table, and there is grief and disbelief when the river turns against the city, smashes down its defences, and lays it waste.

Even the generally placid and genial Trent has always had its unruly side. There was the flood of 1683, when the bridges at Nottingham and Newark were swept away by ice floes driven on by a roaring torrent of floodwater; Candlemas 1795, when thousands in Nottingham were imprisoned in their homes for days, and for six weeks all the land between Lincoln and the river was a lake; December 1910, when a rapid thaw of lying snow combined with days of lashing rain sent floods surging through the slums of Nottingham, overwhelmed trains in cuttings and inundated the floodplain down to Gainsborough.

Nottingham 1875

The last great Trent flood was in the spring of 1947. The river rose to twelve feet above normal at Trent Bridge, Nottingham, and half the city was swamped. Since then enormous sums have been spent on improving the flood defences, and there have been no more disasters. Similarly, in London the Thames Barrier now protects the capital, inducing an easy confidence among its rulers and residents, even though increasingly shrill alarms are being sounded by environmental campaigners about whether it will cope once global warming really gets going. Occasionally, this complacency is pierced, and we are given a sharp reminder that our control over the 'managed environment' is not, and never can be, as complete as we might like to think. In the autumn of 2000 floods caused extensive damage to property and disruption to normal life in and around Lewes and York, and along the Severn valley. The Cornish village of Boscastle was largely wrecked in the course of one day in August 2004 after an eight-inch deluge caused two streams to burst their banks. The city of Carlisle was inundated in January 2005 after the Eden breached flood defences.

Then there were the floods of summer 2007, widely and wrongly described by the BBC and other reputable news outlets as 'the worst ever recorded' or 'the worst in living memory'. Initially they aroused no more than mild interest in London, as they were concentrated on faraway places such as the Hull and Doncaster areas and along the lower section of the Trent. But as sodden June gave way to rain-lashed July, the havoc shifted south and west, along the Severn and Thames valleys. Towns and villages were swamped. Drinking water was contaminated. Families fled and supplies of sandbags gave out. The floods became front-page, top-of-the-bulletin news.

It would be wrong to downplay the impact of serious flooding on human settlements, the misery of those affected, and the economic damage. But there was a time when those who lived along rivers or in floodplains did so in a state of awareness of the precariousness of that existence. They could not have liked the floods when they came any more than did the residents of Tewkesbury in 2007, but at least they were not surprised, and were to a degree prepared. We are not prepared. We have forgotten that, while natural forces can be controlled, they are never subjugated. When major inconvenience comes along – 'disaster' is the media term – the reaction is one of despair and disbelief. Then the sense of outrage sets in, and the search for someone or some organisation – the Environment Agency, the local authority, the government – to blame. 'No one warned us,' we wail.

No one is to blame. Floods occur when the quantity of rain falling from the skies exceeds the capacity of the ground to absorb it and of rivers to take it away. However thorough the attention to river courses and the provision of embankments and other defences, this disparity will occasionally prove too much. If we, as a society, insist on spreading our settlements into vulnerable areas, the responsibility is ours.

Fortunately for us, in our temperate, moderate land, such events occur rarely; and even when they do, their scale – measured in terms of loss of life rather than damage to property – is modest. In other parts of the world, the reminders of nature's capacity to hit back are a lot more forcible. Ask them along the Mississippi about the brown river god of Eliot's *The Dry Salvages*:

> . . . the brown god is almost forgotten
> By the dwellers in cities – ever, however, implacable.
> Keeping his seasons and rages, destroyer, reminder
> Of what men choose to forget.[2]

Seneca would have known exactly what T. S. Eliot was talking about. To him, the river was a many-faceted entity, both provider for us and image of us. While I have no evidence that the high-minded Roman was an angler, I think he would have liked it. I'm sure that, at the very least, he would have understood and approved the notion of the river offering another dimension of possibility; that as well as giving power, a means of transport, sustenance for the crops and food in the form of fish, it would give nourishment for the soul: a particular pastime and relaxation, to 'recreate the spirits'.

You cannot date the evolution of the sport of angling. People devised ways to catch fish because they needed the protein. All methods required skill and knowledge. One – the use of rod, line, hook and bait – tended to be more efficient than net or trap or spear when it came to the bigger, more elusive specimens. Perhaps because it took time and could be done by one man on his own, and because it rewarded cunning and persistence, it was found to give a pleasure over and above the satisfaction of securing food. Long ago they didn't bother with disentangling profit from pleasure, and took both when they could get it. Claudius Aelianus, writing a century and a half after Seneca, described how the Macedonians lured trout to take an artificial fly, which happens to be the most effective way of catching them when they are feeding on insects, as well as being the best possible fun.[3]

In time angling began to challenge hunting as the favourite outdoor pastime in England. The man who did more than anyone else, before or after, to cement the place of the sport in the nation's affections learned to love what he called 'the contemplative man's recreation' here, on the Trent, and on another stream across the fields from where I was sitting with my mug of coffee.

I suppose most half-educated people have heard of Izaak Walton and *The Compleat Angler*, although I doubt if one in a hundred of them has read any of it. It's not easy, but then not much of the prose of the seventeenth century is, and it is surely more accessible than Milton's *Areopagitica*, Burton's *Anatomy of Melancholy*, Ralegh's *History of the World*, Bacon's *Advancement of Learning*, or any of the other literary milestones of the time. To our taste, the conversational form –

Piscator holding forth, Venator prompting, various passers-by butting in – seems comically clumsy. The argument itself is cluttered with deviations, non-sequiturs and homilies on the virtuous life, the observations on fish and how to catch them jumbled with a mass of myth, hearsay, invention and every kind of pseudo-scientific nonsense.

But when he is not being pious, moralistic or ridiculous, Walton is often wonderful. He can be as solid and real as Pepys, with the difference that his theatre of operation is the English countryside rather than the seething city. Walton celebrates meadows and hedgerows and shady trees, as well as rivers, streams, lakes and ponds. He is the bard of good cheer in friendly inns, of banter with milkmaids, of country ways and escape into enchantment. He offers an idyll, rooted in recognisable places and therefore accessible to all. His joy in it comes straight from the warmest of hearts, and it was shaped here, in greenest Staffordshire.

Of course there were fishing books before *The Compleat Angler*, and others before Walton who pondered the magic of the sport. The most celebrated of them, the *Treatysse of Fysshynge with an Angle*, was in circulation well before Caxton's apprentice and disciple, Wynkyn de Worde, published it in 1496. It was subsequently attributed to the pen of a shadowy abbess, Dame Juliana Berners (who may have been a Dom rather than a Dame, and therefore a monk, and have been Julyan, or Juliana, Barnes, Bernes, Berne, or Barne, or may have been someone else entirely). Whoever the Dame was, he or she reached to the heart of the matter:

> I ask this question, what are the means and causes that make a man happy? Truly in my judgement, the answer is recreation and honest pastimes which a man may enjoy without regret. From which it follows that such sports and pleasures give a man a fine, long life. Therefore I will now choose four good and honest sports: hunting, hawking, fishing and fowling. Of these the best, in my simple judgement, is fishing.

In mine too, I might say, and in that of Walton's friend, the diplomat, occasional poet and respected Provost of Eton College, Sir Henry Wotton. Angling was to him, Walton recorded, 'a rest to his mind, a cheerer of his spirits, a diverter of sadness, a calmer of unquiet thoughts, a moderator of passions, a procurer of contentedness . . . It begets habits of peace and patience.'[4]

This is all very true. Elsewhere Walton mused: 'No life is so happy and so pleasant as the life of the well-governed Angler; for when the lawyer is swallowed up with business and the statesman is contriving or preventing plots, then we sit on cowslip banks, hear the birds sing, and possess ourselves in as much quietness as these silver streams which we now see glide so quietly by us.'[5]

Walton was born and brought up in Stafford, and his first silver streams must have been the Sow – which runs through the town – and the Trent, which is no more than a morning's walk away to the north-east. These are reasonable assumptions – Walton disclosed nothing about his early life and very little about the rest of it. His notion of autobiography was to record of his first wife, who bore him seven children: 'Rachel died 1640.' We know that his birth was humble, that he was apprenticed to an ironmonger in London and did well enough to own 'half a shop in Fleet Street'; that his second wife was the half-sister of the celebrated Bishop Ken of Bath and Wells; that his circle of friends included a rich crop of pious divines, public servants, natural scientists and other worthies.

The great cause of his life was the monarchy. Walton viewed the Puritan revolution with revulsion – how, as he put it in his *Life of Bishop Sanderson*, 'the former piety and plain dealing of this now sinful land is turned into cruelty and cunning'. He valued plain dealing above most things, and at some point during the Civil War he left London and went in search of it in the village of Shallowford, a couple of miles north of Stafford. It was a good place for a known supporter of the King to keep his head down: small, out of the way, quiet. It is still small and out of the way, and would be quiet were it not for the main London–Manchester railway line which runs through a cutting forty yards from where Walton's long, half-timbered cottage still stands.

Shallowford had another, potent attraction. Twisting and turning through the fields at the village's edge is a lovely stream, a string of shallows striped with weed, bubbling riffles and dark little pools sucking and circling around the roots of willow trees. Walton remembered the joy of being able to 'loiter long days near Shawford Brook', where he might

> There meditate my time away;
> And angle on; and beg to have
> A quiet passage to a welcome grave.

The passage to his grave was a very long one, but it became quiet enough once England was restored to its senses and a Stuart to the throne. Walton was eighty-three by the time the definitive edition of *The Compleat Angler* appeared, expanded and much enhanced by twelve short chapters on fly-fishing by his young friend Charles Cotton (whom we shall meet again downstream). He was to live another seven years, passing on quietly enough during a savage December frost at his daughter's house in Winchester.

In his will Walton left the means to buy fuel for the poor of Stafford, 'the said coals to be distributed in the last week of January or the first week of February, because I take that time to be the hardest and most pinching times with poor people'. He'd known what it was like to be poor and cold, although he would not have said so. To describe Walton as reticent about himself is like describing Charles II as keen on the opposite sex. The information he supplied about himself would not half fill a wicker creel of modest size. Even with his favourite sport, we know no more than that he fished the Thames at Eton with Henry Wotton; the Lea north of London – because *The Compleat Angler* opens with him stretching his legs up Tottenham Hill, his mind on a brace of trout from that river near the Thatched House at Ware; the Dove in Derbyshire, because Charles Cotton says so; and the brook at Shallowford.

No matter. Walton's magic does not belong to one river. He was not so naïve as to forget that beyond the cowslip banks Royalists and Parliamentarians were slaughtering each other. He was familiar enough with the grimness of the daily grind, the ugliness and nastiness around. But he did not forget, either, that life was a gift and that fishing was one of the ways to honour that gift. He distilled the feeling every fisherman experiences once in a while, of oneness with the river, of the current of the river merging into the current of life, of the exclusion of all else in the moment of the fish taking fly or bait. The ironmonger actually knew a lot about how to catch a fish. But the lasting lesson is how to live.

Bridge ahead

I set off early. We glided through the meadows to Sandon, where the village, like Burston, is some distance from the river. I had originally thought of stopping there for the night, but was deterred by the possibility that the Stafford and Stone Canoe Club might be holding a committee meeting in the Dog and Doublet.

Compared with Burston, Sandon looked rather forlorn. There was a pretty cricket ground, but the outfield was uncut, the sightscreens were on their sides, and the ancient roller had clearly not seen recent action. Across the road was the entrance to Sandon Park, home of the Earls of Harrowby. The closed, spiky gates and the long drive the other side did not invite casual entry. I knew from Pevsner that there was a nineteenth-century pseudo-Jacobean mansion at the end, and that somewhere around was the belvedere originally placed by Barry on top of the tower at Trentham and snapped up at the cash-strapped Duke of Sutherland's sale of fixtures and fittings.

Sandon Hall

An earlier Earl of Harrowby was characterised in Charles Greville's *Diary* as 'pert, rigid and provoking',[6] but Pitt liked him and he liked Pitt, enough to have put up in the south-east corner of the estate a column with an urn on top of it as a memorial to the gouty statesman. The urn pokes out of the trees quite near the river, but I missed it as I was recovering from a slight scare. They have a thoroughly bad habit in these parts – maybe in all parts – of dumping the unwanted old bits of bridge in the river when they build a new one. I thumped the *Otter*'s bottom on a lump of masonry beneath the bridge at Sandon, and again, several times, at Salt, where the river is squeezed between fat, ugly buttresses into two unnecessarily tight arches.

Thereafter we skimmed easily enough to Weston, where I stopped in the hope of having an early pint at the Saracen's Head. Although someone was busy on a stepladder outside the pub, watering the bright bursts of scarlet overflowing from the hanging baskets, it was still shut. I didn't mind too much. The day was too gorgeous for regrets. I followed two teenage girls who were leading a plodding piebald horse into a field by the river. I waved to them and went back to my boat.

We went with the current, and all I had to do was to dip the oars every so often to keep the *Otter* away from the sides. Green fields and thick hedgerows sailed slowly by. Cows munched and waved their tails against the flies. Yellow wagtails swooped from weedbed to tree stump, dipping their heads as they went, at each rest stopping to look quickly round as if to check how their performance was being received. Clouds of black damselflies shimmered over the water. I passed a grove of silver-grey aspens, their foliage rustling restlessly in the breeze. Underneath me the river bed was pale, golden sand, shaped into ridges like the bare mountains of Ethiopia seen from the air.

For a time I felt as if I might have been shunted back to Walton's day. Around the next corner I expected to come upon an angler in breeches and a tall hat, grasping a rod of hazel and blackthorn tipped with whalebone, eyes intent on his line of twisted horsehair dyed in the juice of walnut leaves; or to glimpse an apple-cheeked milkmaid, churns clanking across her shoulders. It startled me when the Sabbath peace was suddenly fractured by the roar of two machines, one slicing the grass, the other whipping it into cylinders of black plastic and rolling them back on to the shorn sward.

The pastoral idyll was then thoroughly terminated at a place called Hoo Mill. The lazy sods had done it again: built a nasty, new, flat bridge of steel and concrete and thrown the old one into the river. I heard it before I saw it, a menacing rumble that grew quickly in volume as we came around a tight bend on to it. With more warning, I would have stopped to inspect the obstacle and dither. As it was, the turn was so sharp that before I really registered what was happening, the current was accelerating and the *Otter* was being sucked towards what looked to me like an extended roaring rapid.

I grabbed the paddle but before I could do anything useful with it, there was a shuddering crunch as we struck the first chunk of masonry and came to a halt. I jumped out and went over my waist in the foaming torrent. The *Otter* was wedged on top of one slab of stone and against another. Then she turned around 180 degrees and became stuck again. Stumbling and slipping, I cursed and heaved. Water sluiced over the stern and

Rough water at Hoo Mill

slopped around my stores and gear. I strained some more and managed to lift and drag the boat through and into a calm glide below the bridge. I baled her out, scrambled in and glanced back. The rapid looked just as fierce and unpleasant from below, the difference being that we were through it. Quiet water lay ahead and my spirits soared.

They were still high when we slid into Great Haywood, which is where the Trent and Stafford's river, the Sow, come together. It is a place treasured by worshippers at the great altar of Tolkien, for it is said to have been the model for the village of Tavrobel in one of his

Sow meets Trent

tales, in which a gnome 'stands nigh the bridge' not far from the House of a Hundred Chimneys.[7] Earnest devotees have debated whether the many-chimneyed residence could be equated with Shugborough Hall, the pale and exquisite seat of the Earls of Lichfield which stands a long stone's throw from the river junction. There is a bridge as well; in fact, just the kind of bridge that would find its way into a rigmarole about elves and gnomes and the ancient days.

Essex Bridge then

Essex Bridge is long, low and remarkably narrow. Its fourteen arches are separated by cutwaters shaped like arrowheads, above each of which is a triangular refuge provided – it is thought – for pedestrians to avoid the packhorses swinging their loads along the four-foot wide passage. The way led, in distant times, between the hunting grounds of Cannock Chase and the great estate of Chartley, which is to the north-east. There is a romantic tale – romantic, therefore probably false – that it was built by Elizabeth I's pet, the wild and self-destructive Robert Devereux, Earl of Essex, so that the Queen could conveniently go in pursuit of the panting hart.

I wondered what Queen Bess and her earl, or the Lichfields or any of the other high-born hunting enthusiasts who came this way over the centuries, would have made of the scene of mass merry-making that greeted me as I approached the Essex Bridge. There were families picnicking and boozing on the banks, boys swinging from trees into the water, fathers guiding sons down the stream on inflated tyres, dogs chasing balls, children splashing in miniature boats, and everywhere pale and reddening skin and over-size bellies on abundant display. Across the bridge itself another stream flowed, of humanity – mothers with pushchairs, elderly couples, trippers, ramblers, strollers, saunterers – taking in the view of Shugborough and the two rivers and the canal and the old cottages.

and now

Three teenage girls in bikinis spotted me. I was up to my knees in the water, guiding the *Otter* over a minor weir. They splashed out to help. One said she had never seen anyone come down the river before. Where had I come from? Where was I going? 'From Stoke? Shit, how long did that take you? All the way to the sea? God, how exciting is that. Can I come? Go on, take us.' I laughed, absurdly pleased to be taken for an intrepid adventurer. I ran the *Otter* on to the gravel beach below the bridge and went off in search of beer, which I found in a black and white timbered pub beyond the canal.

When I got back the afternoon seemed hotter than ever. But the blue was leaking from the sky, as if a layer of polythene sheeting had been stretched between it and the earth. The girls, slim and perfect, waved me off. One said she'd heard on the radio weather forecast that there were thunderstorms on their way, expected that evening. I looked back at them, framed in one of the arches, the two rivers rippling together behind. I turned away to push with the oars. There was tension in the air.

Chapter 8

Rocking Past Rugeley

The character and demeanour of the Trent changed at Essex Bridge. Above the junction with the Sow, you could still call it a stream. Below, it became a river. It was like a boy looking in the mirror and fingering the shadow of the moustache on his upper lip for the first time. It had taken on a new muscularity and physical weight. The profile retained the familiar pattern of shallow, deepening channel, pool, and the water hastened and relaxed in response. But, even with the flow at mid-summer minimum, the scale was different. There was now always at least a foot or two of water under the *Otter*, and I no longer had to jump out and push. In the pools the depths were dark and the bottom beyond the range of an oar thrust.

But it was still very pretty. Along the western bank it was flanked

Shugborough

by Shugborough's park, then by a wood called Haywood Warren; along the east by pastures bounded by the canal. If I stood I could see the pots and baskets of flowers on top of the narrowboats, stationary beside the towpath or creeping along like brightly coloured beetles following a favourite path.

A waterside pub came into view. Sun-basted torsos were widely distributed across the gardens, from which dogs plunged into the water in pursuit of balls and sticks past a sign warning humans not to swim or fish. I floated past, my mind for once on something other than beer. The sunshine had gone and the sky had been painted with a grey, metallic sheen which boded ill. The girl's warning about thunderstorms stayed with me. In the distance, over the massed ranks of conifers standing guard on Cannock Chase, there were mutterings in the atmosphere, like a distant argument.

I glided through the shadowy recess under the stone bridge taking the A51 over the river. The canal came close again, then the river cut away west as if seeking privacy. There was a line of alder bushes on the right, and on the left a little inlet with a shingle shelf. I ran the *Otter* aground and peered over the bank. There was a meadow with sheep in it, and on the far side a hump-backed, red-brick bridge over the canal, with a road beyond. I pulled the boat up as high as I could, unshipped my camping gear and stores, swept a patch of grass clear of sheep droppings, and had my tent up in twenty minutes.

Bishton camp site

Overhead the argument was getting closer and louder. A drift of fat raindrops pattered against the outside of the tent. In urgent haste I blew up my mattress, gathered my stuff into the little porch and got the blue cover fastened over the *Otter*. I scurried inside the tent, zipped it close and lay on my back. Although it was only twenty past six on the second longest day of the year, it was too dark to read. Anyway, it was too noisy. There was a succession of tremendous crashes, as if howitzers were being fired in the next field or an enormous bass drum was being beaten with maximum force. Looking up, I could follow the flickering of the lightning through the fabric of the tent. The rattling of rain became more insistent, its pitch deepening to a muted roar. I pictured the dry, hot land lifting up its arms to receive the water.

We don't often get that close to the violence latent in the skies. The clouds generally hold their peace and their distant stations, remote and self-contained, going about their business quite independent of us. They appear as weightless shapes in the air, rather than being made of the air and heavy with water.

Clouds are vapour made visible. The water comes from several sources. It is raised from the sea and the lakes and rivers. It is evaporated from wet ground and vegetation. It is exhaled through microscopic pores in the leaves of trees and plants, having been sucked by them from the ground through their roots. The quantities can be surprising − a hectare of rain-drenched forest can return ten tons of water in an hour to the air above.

The water molecules rise invisibly on currents of warm air. As they encounter bands of cooler air, they come together into droplets and take visible form. When the rate of condensation is greater than that of evaporation, clouds are made. When the droplets become big enough and heavy enough to overcome the upward air draught, rain falls.

Cumulus is the billowing cloud of our summer days. It is born as the sunshine warms the ground and the layer of air above it, which then rises with its load of vapour. It is mostly peaceful and benign, capable of producing a light shower, not much more. It has a cousin, though, much more excitable.

If the air is very moist and the ground temperature is very high and the updraughts are cooled very rapidly, a prodigious, self-fuelling tropospheric pandemonium is generated. At the summit, as much as 60,000 feet up, ice crystals are spread flat to form the thunder-cloud's distinctive anvil top. Through the middle is a column, the engine of the cloud's energy. Colossal charges of electricity leap across and down it, tens of millions of volts released in millionths of a second. Globules of rain are flung upwards by winds powerful enough to shred an aircraft, flung down again, then up, time and again until they are discharged earthwards as hail. No wonder that the thunder should have been taken for the voice of a displeased god. In the words of Sir Robert Grant's great hymn, *O Worship The King*:

> His chariots of wrath/The deep thunder-clouds form,
> And dark is his path/On the wings of the storm.

Before, with, or after the thunder comes the rain: not Portia's gentle dew, but sheets, buckets, stair rods, cats and dogs. The volume can be too much for the ground to deal with. Lawns become ponds, sports grounds lakes, ditches foaming streams. The strong, musty smell of wet dust rises as the water churns down the sides of roads, backing up as it tries to force its way down over-loaded drains. It has only one aim: to seek the river. And the river quickens, swells, colours and becomes menacing.

I lay and listened to the storm. In time the drumrolls of thunder receded, and I poked my head out to take a look. Heavy rain was still falling, and I was surprised and pleased that my tent had stayed dry inside and that the pegs had held in the darkened earth. The cover on the *Otter* was bowed under the weight of a pool of water.

Intending to scoop it off, I removed all my clothes except my boxer shorts and took a step on to the muddy slope leading down to the shingle ledge. My foot shot out from under me and I slid down sideways, mud clinging to my backside, legs and elbows, swearing in the usual pointless fashion.

The rain stopped after an hour and a half. The meadow was left draped in a quilt of mist, through which the grub-like shapes of the sheep inched their way nibbling as they went. Water dripped in beads from the tent ropes and cascaded from the alders along the far bank. I heated baked beans and spam and ate the mush straight from the frying pan, followed by a slab of the fruit cake my wife had made for me. I drank two or three glasses of red wine while making some notes, then read Magris' *Danube* until it was too dark to distinguish the words. The passage was about Budapest and ended with an account of a gypsy violinist playing a tune called 'The Skylark' in a celebrated restaurant, the Matthias. I had been there many years before, and there had been a violinist – maybe the same one – with greasy black hair and flashing smile, squeezing out one sentimental melody after another. I resentfully pictured the great polymath returning to his hotel by the river, sliding unthinkingly between crisp sheets. I retreated, moist and smelling of mud, into my sleeping bag.

Morning brought the smile back to the land and the skies. The storm front had marched north, leaving the air cleansed and sparkling. The sun spread a pale fire as it rose over the canal, consuming the rags of mist. The blue overhead promised a perfect June day. But the river was a different proposition. It was up two feet, maybe more, enough for the *Otter* to be afloat and for me to bless my unusual good sense in placing the mooring stone high up on the grass. The water was charged with sediment, the colour of oxtail soup and with something of its texture. It was at work, moving earth.

*   *   *

It's what rivers do, shift soil. Simultaneously and ceaselessly they convey and deposit.[1] They erode the landscape and reshape it, sometimes in a night. It is astounding how one big flood can change everything. Huge boulders are rolled away, islands and sand bars shrink, change shape, expand, disappear altogether. The shape of a pool is compressed or distorted, its neck jumps sideways, its tail migrates from one side of the river towards the other. Familiar marks are buried, long-buried treasures are revealed.

The soil was all rock once, and water on its own cannot erode rock, or anything else. The wearing away is done by the abrasive particles suspended in the water – sand, grit, fragments of stone that have been picked up and carried along. The more powerful the flow, the bigger the particles are and the greater their impact. Even so, the river's force is nowhere near great enough to shift a boulder, yet shifted they are. The trick is to attack the gravel and sand supporting the boulder. Wash that away and gravity takes over, rolling it down to its next resting place, where it may remain for a minute, a month, or five hundred years.

It stands to reason that soil – old rock – is moved much more easily than bits of new rock. Less obviously, medium-sized particles are more readily shifted than finer, lighter ones, which tend to settle closest to the stream bed, out of the reach of the strongest flow and held together by electrical forces. The overall sediment load is made up of all sizes, with the biggest fragments being rolled and bounced along the bottom. This carrying capacity is enormously expanded in a flood, but even then the river deposits as it goes.

This process sculpts the river bed into ripples and dunes which, over time, migrate downstream. A swift flow over coarse sand deposits visible, highly unstable bars. Sometimes these are repeatedly washed away only to re-form, splitting the river into a multiplicity of channels separated by shifting islands. On lowland rivers the architecture is less dramatic. Even so, season by season, great changes take place.

All sediment has to come to a stop somewhere. The bigger particles drop out early, wherever there is a momentary slackening. The smaller ones usually find their resting place along the flood-plain, where the gradient flattens out. Over the ages of sediment-dumping, the bed of the river has been lifted. Each time the river has flooded, a layer of mud has been laid down on the plain. Left to itself, the river will forever be cutting new channels across that plain, meandering left and right, east and west, north and south, forever reinventing itself. The job of engineers and dredgers is to check that digressive tendency.

The finest stuff is held in suspension the longest. As the river meets the sea and slows, the salt prompts the clay particles to cling together and sink. In the case of river mouths exposed to powerful tides and currents, the sediment is dispersed and an open gulf or estuary is maintained. But where the movement of the sea is comparatively sluggish, a delta is created: a mass of silt split into a fan-like web of channels through which the water has to find its way. Over time the silt hardens into land, and the sea is pushed back, which is what happened with Miletus, where Thales did his thinking.

The river surging past me was transporting more than the reddish earth of Staffordshire. Branches scooted by, a tyre, a cushion, a football, a mass of reeds and sticks and weed. I listened to the news on the radio as I drank my coffee. The lead item was a report from north Yorkshire, where floods had burst through villages, upended cars and marooned livestock. Over my second cup I stared at the Trent and tried to picture the *Otter* afloat on it and me in the *Otter*. It was not a comfortable image. There was a sudden commotion beneath the bank. A waterlogged sheep was trying to find a purchase in the mud for its sharp little hooves. After a struggle it managed to force its way on to the land, where it staggered drunkenly under the weight of its fleece. It stood for a while, water sluicing from it, then shook itself and trotted off to find new friends.

I decided to leave the river until it had calmed down. A day with the firm earth of England under my feet beckoned. I examined my map. A minor road led north towards a patch of blue called Blithfield Reservoir. It looked promising, so I took it with the vague idea of crossing the water and reaching the village of Abbots Bromley. This is famous for its Horn Dance in early September, when some of the residents don medieval costumes and caper around brandishing sets of deer antlers. The origins of the rite are enormously ancient and have been picked over assiduously by historians, folklorists, New Agers and other enthusiasts for our primitive past. Some maintain it commemorates the granting by Henry III of grazing rights in Needwood Forest; others that it belongs to dimmer Druidical times when pagan worshippers would dress up in animal skins or horns in the hope of promoting success in hunting; others that it is a fifteenth-century morris with elements of Robin Hood, Maid Marian and the hobby horse thrown in.

Wherever the truth lies, my desire to investigate it waned as I ascended the long hill under an increasingly brilliant sun. At its crest I looked down into a hollow partially filled by the blue waters of the reservoir. A causeway curved across it carrying the road to Abbots Bromley. I could see the steeple of the church and the rooftops, but they seemed a long way off and it was hot and getting hotter. I paused for a while in the car park, where there was a notice announcing that one and a quarter million people received their water from the reservoir; and that the resident birds included swallow, swift, mallard, Canada goose, swan and cormorant – none of which struck me as much to boast about.

This whole district was once the property of a family called the Bagots.[2] They were a very old family, possibly as old as the Abbots Bromley Horn Dance, depending on whose dating you favour. I liked the sound of Bagot: a solid, well-rooted sort of name. Their estates have long since been broken up, and they themselves have been dispersed far away. But along the western shore of the reservoir, beyond an unappealing development of wooden holiday chalets, the high chimneys of their ancestral home, Blithfield Hall, appeared above a screen of oaks. There, through the grounds and across the gardens and between the trees, a creature roamed of which the family were as proud as they were of their own name and lineage.

Blithfield Hall

The story goes that the first Bagot goats — a breeding pair, one presumes — were a gift from Richard II to Sir Ralph Bagot to show his appreciation for an exceptional day's hunting in Needwood Forest (another version is that they were souvenirs brought home by returning Crusaders who had found them somewhere in France). For the best part of six hundred years their descendants chewed their way around Bagot land, wild and wary, valueless for meat or milk, but admired for their black heads and chests, their white backs and the elegant sweep of their horns. Although the line has been preserved elsewhere, there are apparently no longer any Bagot goats outside Blithfield Hall, just as there are no Bagots inside.

I walked back up the slope from the reservoir to Admaston, a hamlet of five houses, where I was hoping to ask someone for a drink of water. There was not a soul about. I stopped by a sign which read: NO TURNING NO PARKING NO DOG FOULING — PICK IT UP.

Admaston is 400 feet up, which is high for Staffordshire. I looked across to the slaty slope of Cannock Chase. To the left the cooling towers of Rugeley Power Station steamed noiselessly into the blue sky. The canal and the Trent were hidden from view, but I could see clearly how the landscape had moulded the watershed. The river was squeezed against the escarpment of Cannock Chase, while on the near side the terrain opened into a descending succession of folds and wrinkles, each pointing at an angle to the river to offer its contribution.

I skirted the southern edge of what was left of the Blithfield estate and took a path downhill. At the bottom I crossed the Bourne Brook, a little ribbon of water concealed within a band of trees and bushes. The land rose quite steeply the other side. Beyond the grazing cattle I spotted a bull, his enormous, dangling testicles visible from the far side of the field, along which I watchfully hastened. At the top of this hill was an empty red-brick farmhouse with holes showing in its red-tiled roof. A high-banked lane took me up, then down, then up, then down again towards the river. Light-headed with hunger and thirst, I marched with sore feet across the stone bridge and into the pub I'd passed in the *Otter* the previous afternoon, the Wolseley Arms.

The story of the Wolseleys goes as far back as that of the Bagots, maybe even further. The name is said to come from Wolves-ley, the family having been granted their lands in return for exterminating the wolves that once infested 'Cranke' (Cannock) Chase. While the Bagots were generally for King, country and a quiet life, a streak of non-conformity ran through the Wolseleys. When the Civil War broke out, the 1st baronet, Sir Robert, declared for Charles I. But his eldest son, Charles, was a staunch Cromwellian; as was another son, William, who followed a soldiering career and found himself in trouble with James II for having the Mayor of Scarborough tossed in a blanket for 'indignities inflicted on Protestant clergymen'. The 7th baronet, Sir Charles, was in Paris in 1789 to cheer on the storming of the Bastille and became such a radical that he eventually served eighteen months in Abingdon prison for sedition.

Wolseley Hall

The seat of the Staffordshire Wolseleys, Wolseley Hall, stood close to the south bank of the Trent. Three centuries ago that tireless tourist Celia Fiennes stayed there for some weeks at the invitation of her aunt, who was married to Sir Charles. 'His seate', she noted in her somewhat disjointed way,

> stands very finely by the river Trent, there is also a moate almost round the house; the house is old timber building, only a large parlour and noble staircase with handsome chambers Sir Charles has new built . . . It's a large parke 6 miles round full of stately woods and replenish'd with red and fallow deer, one part of it is pretty full of billberyes which thrives under the shade of the oakes, it's a black berye as big as a large pea and are ripe about harvest; there is a very ill custome amongst them, now not to be broken, when they are ripe the Country comes and makes boothes and a sort of faire the outside of the parke and so gather the berries and sell them about the Country; the greenes they call Wissums and these Wissums the deer browse in the winter and on hollye of which there is a great quantitys . . .[3]

The land where the bilberries grew and the deer browsed is now occupied by a Wyevale garden centre. For a family connection I had to make do with the Wolseley Arms, motto: 'Inn Keeping With Tradition'. While I sat waiting for my bangers, mash and red onion gravy, I inspected a medley of wise sayings inscribed on blackboards and hung about the interior. 'To do nothing and get something formed a boy's ideal of a manly career' – DISRAELI. 'Power corrupts and absolute power corrupts absolutely' – ACTON. 'To do what you like is freedom. To like what you do is happiness' – PAT. I worked out that Pat was the tough-looking blonde on the other side of the bar, who was taking orders, instructing staff and exchanging banalities with the clientele in one seamless, efficient display. Under the awning outside, two middle-aged couples – one local, one visiting French – were lunching together. The two men said almost nothing, while the Englishwoman maintained a gallant but halting discourse about the weather and features of the district in evening-class French. The waitress brought the French woman a piece of fish in batter as big as the blade of a cricket bat. '*C'est incroyablement énorme,*' she said, awestruck.

Bishton Hall

After lunch I plodded back over the bridge and down a lane past a large grey Georgian house, Bishton Hall. It was built by a banker, a Mr Sparrow, and would have looked across the river to Wolseley Hall. Without any justification, I pictured Mr Sparrow as a bit of a parvenu, and the Wolseleys – still the squires, whatever their politics – regarding his intrusion with much huffing and puffing. Bishton Hall is now St Bede's Preparatory School. Across the lane was the school cricket field, flanked along one boundary by beeches and with a fine, spreading chestnut at wide midwicket. There was a little black scoreboard beside the wooden pavilion on which some young wag had put up an absurdity: 303 for 5, last man 292.

Bishton cricket ground

I reached my meadow, checked the *Otter* was all right and stretched out in the shade of the hedge along the canal. When I woke up I spotted a gang of teenagers approaching from the downstream end of the field. When they reached the river most of them jumped in. I went over to the boat, ready to repel boarders. A youth wandered up the bank towards me accompanied by a girl in a black bikini. He had on long swimming trunks and his hair was dyed pink and blond. He told me he was a punk rocker and played in a band in Rugeley. He told me its name but I was distracted by the sight of the girl, who was carrying a good deal of metal around and in her nose, ears, mouth and belly button, and I forgot it. I asked him if punk was flourishing in Rugeley. He laughed scornfully. I asked him what Rugeley was like.

'Rugeley,' he said carefully, 'is full of wankers.'

Chapter 9

The Ale of England

I never got an opportunity to test the punk rocker's assertion. Although the canal – having crossed the Trent on an aquaduct – cuts through Rugeley, the river itself meanders past the town to the north-east. I regretted not being able to check on the incidence of wankers. I should also have liked to visit the Talbot Arms, the scene of a satisfyingly grisly episode in the grisly career of Rugeley's most famous son – the Victorian poisoner William Palmer. It was in an upstairs room in this hostelry that they laid the exhumed coffin of one of Palmer's alleged victims, his brother Walter. When the coffin was opened, the inquest jury summoned for the occasion was so overcome by the stench that several of them vomited on the floor. Those who dared look in reeled back from Walter's hideously swollen face and distended limbs – 'a mass of dropsy, corruption and gangrene', one observer recorded with relish.

They never managed to pin the murder of Walter on Palmer; or, indeed, those of his mother, wife, several of his children and at least one of his many girlfriends, all of whom met sudden and unexpected deaths in his vicinity. The one charge of which he was convicted, and for which he was hanged – that of the murder of his friend John Parsons Cook – was sustained wholly by circumstantial evidence. Palmer's final words, addressed to his solicitor – 'I am innocent of poisoning Cook by strychnine' – have been interpreted as suggesting that he might have used something else. His last flourish was to present his lawyer with a slim, improving book entitled *The Sinner's Friend*.

There was no missing Rugeley's other main claim to outside attention, its power station. When I emerged from my dewy tent soon after five o'clock the next morning, the first sight to greet me was its great towers steaming silently. I checked the knife that I had stuck in the mud to mark the high point of the flood. The water was still coloured but its level had fallen by at least a foot and it was navigable. We were away soon after six, slipping down-stream at a good pace without any need to pull on oars. Caddises were breaking the surface, skidding in erratic jerks towards the bank. Wagtails were working the margins and the swallows were swooping and diving. I left the sheep to their nibbling and passed under the canal, then a road bridge, then a rusting railway bridge festooned with weeds and outbursts of bracken.

As I got closer, I could hear the steady panting and grumbling of the power station. From a distance these mighty installations have a kind of remote, monumental beauty. The clean, concave lines of the cooling towers and the soaring, slender chimneys magnetise the eye. Exhalations of steam hang over them, replenished myste-riously from below. Across the landscape march the pylons, like skeletal warriors approaching their masters for orders, or departing on their missions. The cables loop to every point on the compass, and you have the feeling that you may have stumbled across a megalomaniac scheme for global domination.

Close at hand they are less impressive. Clinging to the skirts of Rugeley's towers was a clutter of scruffy, grubby ancillary buildings with a thicket of transformers at ground level. Of the cooling towers, two were terracotta in colour, and the other two much paler. All of them were smeared and smudged with grime. Beside them rose a slender chimney, coral with a white tip like a cigarette with paper and filter quirkily exchanged. A trick of the morning light converted the steam from white to purple-black against the blue of the sky.

In their time the power stations represented a solid and unquestioning confidence in Britain's industrial future, and nowhere was that confidence more loudly proclaimed than along the course of the Trent. The proximity of the Midland coalfields and the availability of water for the cooling towers encouraged the building of thirteen generating complexes along the river. They produced a quarter of the nation's electricity; it was known in the industry as Megawatt Valley.

In the case of Rugeley, two nationalised mastodons – the Central Electricity Generating Board and the National Coal Board – came together in an epic embrace. Rugeley A opened in 1963, fuelled by coal from its own colliery, cooled by the Trent. Rugeley B, bigger by far, came a few years later, and for a few years they steamed away together. But the day of the mastodons was coming to an end, and the belief in Britain's industrial future was crumbling fast. The colliery closed and Rugeley A was demolished. Other monuments of Megawatt Valley suffered the same fate, while some were converted to gas, some stuck with coal, and some were just abandoned.

The power station speaks of man's ascendancy over the landscape. It is a visual statement that we are in charge, and we do as we see fit.

In one way or another, water has always been power. More than three thousand years ago, China's rulers moulded the first hydraulic society, commissioning canals to bring the water where it was needed, and defences to keep unruly rivers in order. Attitudes to water informed a philosophical debate. In the *Tao Te Ching* – the source of Taoism – water is presented as a model for human conduct by giving way to obstacles it cannot overcome but, in the end, wearing down opposition. It incorporates *Wu Wei*, the idea of moving with the flow and with the flux of Nature; the political moral is that leaders should employ minimum force in governing the people. A Taoist engineer of the sixth century BC, Chia Jang, said: 'Those who are good at controlling water give it the best opportunity to flow away, those who are good at controlling the people give them plenty of chance to talk.'

Needless to say, the rulers favoured the more authoritarian tendency of Confucianism, which placed respect for status and active pursuit of virtue above obedience to the ways of nature. It decreed that rivers needed to be disciplined and made to assist the common good. The elite adopted this precept with enthusiasm; the centralised control of the system of water engineering and transport was one of the crucial factors – some have argued *the* crucial factor – in the establishment and maintenance of China's imperial power.[1]

Other empires took the same watery path. The fabled Semiramis of Assyria, legendary founder of both Babylon and Nineveh, is said to have boasted: 'I have constrained the mighty river [the Tigris] to flow according to my will and let its waters fertilise lands that had been barren.' Her flesh and blood successor, Sennacherib, made Nineveh his capital and ordered the construction of an elaborate network of canals to keep its parks and gardens watered, and to develop agriculture across its hinterland. Egypt's dominance of the pre-Roman ancient world was directly linked with the twin stripes of green and fertile land either side of the Nile on its journey through an otherwise desolate waste of sand, and with the network of irrigation canals and storage basins fed by its never-failing waters.

The European civilisations, starting with the Greek and Roman, did not depend to the same degree on irrigation, because they were able to rely far more than desert cultures on rainfall. Even so, there was still the same desire to bend the waterways to the needs of society, to force the rivers to fulfil a useful, or at least non-disruptive, role in the managed landscape. The preferred model, made possible by technology, was to confine the river within a single channel with raised banks or levees to guard against flooding.

The natural order was replaced by our order, and one of its assumptions was that the power of water should be at our disposal.

Between the river and the spiky fence protecting the Rugeley Power Station and its secrets, someone had created a cheerful little golf course. Red flags fluttered over the shorn, dewy greens as I glided by. On the opposite bank cattle chewed around the feet of the pylons.

We went under another railway bridge, this one taking the line south-east towards Rugby. On the other side of it a woman in green overalls and black wellingtons was leading a herd of Holstein/Friesians from their sheds towards the field. I waved a greeting to her, but she was concentrating on keeping their black and white backsides on the move, and she ignored me. I skirted to the south of three pleasingly named hamlets, Hill Ridware, Mavesyn Ridware and Pipe Ridware, the common element being derived from 'rhyd-ware', the Anglo-Saxon for 'river people'.

According to Stebbing Shaw's indispensable *History and Antiquities of Staffordshire*, the north side of the river at Mavesyn Ridware was exceptionally rich in fish, perhaps because 'within the boundaries of this fishery there is an unusual number of deeps and shallows so necessary to the different tribes with which it is plentifully stored'. Among the tribes were pike, perch, grayling, eel, gudgeon and crayfish 'in plenty'; barbel and chub 'in large shoals'; carp 'very rare'; and 'within memory a brace or two of salmon but these were white and out of season'. Best of all were the trout − 'few . . . but those few the pride of the river . . . they cut as red and carve as firmly as the best Newcastle salmon' − and the burbot, referred to by Shaw as 'highly esteemed but a rarity here'. More than a rarity now − the burbot was officially declared extinct in England some years ago, although every so often there is a report in the angling press of a specimen supposedly being landed from some murky hole.

Poor Stebbing Shaw! He was of a delicate constitution, and his labours on his history and the worries arising from preparing it for publication apparently sent him mad, leading to his death in 1802 at the age of forty. I think he would have been pleased to see the condition of the river two hundred years later, and I would have liked to have followed his recommendation and trotted a float and a juicy worm through one or two of the slow, eddying pools that lay below the weedy shallows. But I was conscious of being a day behind because of the flood; and anyway, in such high summer weather, you need to be fishing as dawn breaks or dusk creeps in to have much of a chance. So I urged the *Otter* on past the high banks, shaggy with nettles, meadow sweet, loosestrife and eruptions of the dreaded Japanese knotweed.

I passed the mouth of the Blythe at the end of its journey from Blithfield Reservoir without even noticing it. Then, needing to stretch my legs, I beached above the handsome but redundant pink and khaki stone bridge at Yoxall, beyond which a horrible new concrete bridge showed its ugly face. As I strode through the meadow to the road, I disturbed clouds of damselflies with black wings and aquamarine bodies that scattered in iridescent alarm. I walked up to Yoxall and bought some chocolate and a newspaper at the village shop, then walked back again. Below Yoxall the river cut through more green and deeply peaceful pastureland. Over to the left I glimpsed an imposing red-brick building through a screen of trees.

This was Wychnor Park, now a country club. The earlier manor house and the lands around were held in medieval times by a family called Somerville, and Wychnor was known for a curious custom.[2] According to the chronicles, whichever Somerville was lord of the manor was expected to keep a flitch of bacon hanging in the hall at all times except Lent, 'that it may be delivered to any man or woman who shall come and claim it and at the same time swear that he or she has been married for a year and a day without quarrelling or repenting'. Two neighbours were required to back up the deposition. If the successful claimant was a freeman he received, besides the bacon, half a quarter of wheat and a cheese; if a villein, half a quarter of rye. These items, with the bacon, were 'carried before him with trumpets, tabernets, minstrels and a procession of the tenantry though the lordship of Wychnor and then, without music, to his abode'.

It is a comment on the elusiveness of complete marital harmony that, even seven hundred years ago, claimants for the Wychnor flitch were few and far between. One couple who managed to lay their hands on it immediately began arguing about how it should be cooked, and were forced to give it back. Another were a mariner and his wife, who had not seen each other between the day of their marriage and the day they brought home the bacon. The third were 'a simple pair in the neighbourhood, the husband a good-natured sensible man and the wife luckily dumb'.

At Wychnor, the Trent and the canal that has kept it company since Stoke come wholly together for the only time. For a distance of about 200 yards they are one and the same. Then they change places, the canal heading in a straight line for Burton, the river wriggling carelessly off to the east. This meeting place is close to the village of Alrewas, which is full of picturesque, half-timbered cottages and houses and is said to have been the first place in Britain to be supplied with piped North Sea gas: an interesting and little-known fact.

Just after saying farewell to the canal, the Trent rushes over its first proper weir. My friend at the Environment Agency, whom I had consulted about matters to do with fish and navigation, seemed to think I would be able to shoot it in the *Otter*. I disagreed. The smooth, quickening approach, the foaming drop, the whirl of currents below – all struck me as being thoroughly hazardous. So I dragged the *Otter* out of the water and along the towpath and relaunched her in a quiet eddy well away from the white water.

We floated down a long streamy shallow, braided by beds of weed. Little pale watery ephemerids hatched and were snaffled by damselflies. Two railway bridges passed overhead, followed immediately by the arrival of the River Tame.

As recently as thirty-five years ago the union with the Tame spelled death to the Trent. 'A vicious river,' Peter Lord called it in his *Portrait of the Trent*. When he was writing — his book appeared in 1968 — the volume of the Tame was slightly greater than that of the Trent; and of that discharge, one third — 70 million gallons a day — was untreated sewage and industrial effluent. The Tame flowed out of the heart of what used to be called the Black Country. It was, and had been for a very long time, biologically defunct. It was so toxic that for at least eight miles below the confluence there were no fish at all in the Trent and precious little life of any kind.

In Walton's day the river was famous for its fish. Its name was said — erroneously — to be derived from the French for the number of species it could boast. Shakespeare's friend Michael Drayton listed them in florid style in his poetical ramble around Britain, *Poly-Olbion* — 'the perch with pricking fins . . . the tyrant pike . . . the trout by Nature marked with many a crimson spot . . . the lusty salmon . . . the grayling whose great spawn is big as any peas . . . the dainty gudgeon . . . the flounder smooth and flat' etc.[3]

Eels were trapped, salmon were netted, anglers rejoiced. But early in the nineteenth century the runs of salmon began to decline drastically because of pollution and obstructions to the spawning grounds in the form of weirs and locks. By the 1880s the Trent was poisoned near its source by filth from Stoke, in its middle reaches by the brewing waste and sewage from Burton, further down by Nottingham's mines and textile industries, and towards its tidal section by effluent from the fertiliser plants and breweries of Newark. Among its main tributaries, the Tame contributed the muck from Tamworth and Birmingham, the Churnet (via the lower Dove) the sludge from the copper works and dye factories of Leek and Froghall, and the Derwent the discharges from the Derbyshire cheesemakers and the cloth and silk mills of Derby.

No river system could survive such treatment. Salmon catches went from bad to worse to non-existent. A map of 1937 attached to a report from the Trent Fishery Board's clerk and biologist, J. Inglis Spicer, shows 90 per cent of the river between Stoke and Burton either incapable of sustaining fish or incapable of sustaining life at all. Despite Mr Inglis Spicer's heartening references to the 'vast improvements in sewage disposal' in the Potteries area, and a general 'definite, if slow, recovery', the situation was even worse by the 1950s. The river was entirely fishless from Stoke down to Great Haywood, and largely so down to Shardlow (above Nottingham), and again from Gainsborough along the tidal stretch.

Somewhere along the march of progress we lost our reverence for rivers. River gods and river spirits fell from favour. Instead of seeking accommodations with the natural world, we strove to achieve domination. As a result that world ceased to have an intrinsic value of its own. Its only use was to serve us, and we became antagonistic to those features of it – mountains, for instance, and deserts – that we could not control and exploit. Under the utilitarian ideology, no landscape had a case to remain untouched. Nothing warranted protection or preservation because it was beautiful or extraordinary. The single question was: can it be made to serve? If so, serve it must.

Technology made us masters of most of what we surveyed. Four hundred years ago, across Europe, hydraulic engineers were channelling and shortening rivers, dredging, building embankments, draining wetlands. Later came the age of canals and water transport; later still the age of greater canals connecting seas, like Suez and Panama. The vision of the engineered planet inspired capitalist and socialist dreamers alike, and the concrete dam became the great symbol of our power over landscape. The Americans dammed the Colorado and turned the desert into fields. Stalin harnessed the energy of the Volga, the Dnieper, the Don and a host of lesser rivers to power Russian industry and bring light to millions of Russian homes.[4]

Rivers could not resist their enslavement, and inevitably they forfeited respect as we confined them and bent them to our will. They were reduced to being sources of free energy, means of transport, conduits for our rubbish. In *Water and Dreams*, Gaston Bachelard suggests that the essence of water – its cleanness and purity – acts as an invitation to an unconscious drive, which he characterises as 'a summons to pollution', a longing to commit 'an outrage against nature, the mother'.[5]

Maybe. Certainly England's industrial pioneers could not have cared less about the impact of their ventures on the landscape. Creation of wealth was their mission, and if that required the land to be despoiled and the rivers to be fouled, so be it. The notion of care for the planet meant nothing. One suspects that even if the argument had been carefully laid out, they would have replied that the rivers must have been created for a purpose otherwise they wouldn't be there, and that the purpose identified by society was to remove society's unwanted mess. The extinction of life in them was therefore a price well worth paying.

In the course of the twentieth century, that complete indifference began to fall out of fashion. Concern for clean water revived – not from any great desire to rediscover its sacred nature, but because of the correlation with public health. Water was still a commodity, just a rather more valuable one, whether used for drinking, making electricity or growing crops.

But perhaps we have moved on again. For one thing, we have lost some of that bright confidence in our ability to manage the earth for our own benefit. The mighty engineering projects – particularly dams and irrigation schemes – that half a century ago promised progress and prosperity for all are increasingly seen as symbols of hubris, monstrously expensive and environmentally devastating. The planet itself has revealed in global warming a capacity for exacting retribution that was undreamed of a generation ago.

At the same time, across Europe and, to a degree, the United States, the industrial landscape has been disappearing with amazing speed. We have a tendency to sigh sentimentally over the fate of coal-mining, car-making, ship-building, steel-blasting, copper-smelting and the rest of it; to lament the loss of 'close-knit communities' and irreplaceable craft skills. But there is a dividend for which we might be more appreciative than we are.

Even as Peter Lord was railing against the 'vicious' River Tame, a long-running programme of improved waste treatment works was having a dramatic impact on water quality throughout the Trent system. Since then the collapse of much of the industry on which the prosperity of the West Midlands was based has turned improvement into rebirth. Much the same has happened across the country. Fifty years ago the estuary of the Tyne – which had once produced 100,000 salmon a year – was biologically dead and not a salmon could get through. Now, once again, it is the most prolific salmon river in England and one of the best in Europe. When the Queen was crowned in 1953 the Thames in London sustained no life. Now there are fifty species of fish swimming through the capital, and the river as whole is as thriving an eco-system as when Izaak Walton walked up Tottenham Hill.

We should rejoice over this river renewal. We don't, because we're too busy worrying ourselves into a frenzy over global warming; and anyway very few of us have anything intimate to do with flowing water any more. But I am delighted, and I wish to place it on record that I regard being able to catch a roach from the Tame rather than a turd as a considerable improvement.

With the addition of the benign waters of the Tame, the Trent becomes a considerable river. Through the deeper sections the current slowed to almost nothing, and I had to row hard to keep us going. But each dull, slow stretch was succeeded by a streamy, weedy shallow, and I would swivel the *Otter* around so that I could choose which channel between the weedbeds to take.

As the afternoon wore on, the skyline ahead filled with the towers and chimneys of Drakelow Power Station. Built in the 1960s, this once produced more electricity than any plant in Europe. But heydays in the world of power generation are short-lived. Drakelow became defunct a few years ago, and, since I floated past that day, the towers that gave it its useless grandeur have been blown up. To make way for the power plant, a mansion that had stood since Elizabethan times was bulldozed, and gardens and grounds that had graced the Trent for centuries were levelled and covered in concrete. An old photograph shows a tall-chimneyed pile wreathed in Virginia creeper, with terraces and lawns and borders and wide stone staircases sweeping down to a riverside balustrade adorned with urns. The frontage is now thick with undergrowth in which are clasped the decaying remains of the pumping station and the sluice gates, encrusted in mud and dried weed.

The Trent executes a great loop around the north of Drakelow. The effect was that – having passed the towers and chimneys and left them behind – I found they had reappeared in front of me, giving the illusion of time moving in the wrong direction. I was reflecting on this mystery when I was rudely distracted by an attack from an insanely aggressive swan. This horrible bird evidently resented the arrival of the *Otter* on its territory and launched itself across the water at us, beating its wings, hissing malignantly and thrusting forward its orange bill. There is a myth that swans can break human bones with a blow from a wing. I knew well enough that it was a myth but if ever there was a swan intent on proving me wrong, this was the one. I grabbed the oars and heaved hard. We retreated ingloriously in the direction of Burton, leaving the bird in command of the river.

It was evening, a still, glorious summer's evening, when we glided in view of the town. After three nights in the tent and a long, sweaty day at the oars, my thoughts were concentrated on obtaining hot water, clean linen and a yielding mattress. I stopped first at a riverside pub where the barman said I was welcome to moor but he wouldn't advise it because the area was, as he put it, rough. He suggested trying one of the rowing clubs further down. I was pulling hard for the first of them when there was a crash behind me and a jolting impact that threw me to my knees. I looked back. The *Otter* was undamaged, but the rower in the scull that had hit us was doubled up over his oars, gasping with pain, while his coach berated him from a distance for being on the wrong side of the water.

Upstream from Burton Bridge

I continued more cautiously to the Leander Club where I was introduced to the captain, a strapping young fellow in shorts and T-shirt, bronzed, muscled and bursting with vitality. He said I was welcome to leave the *Otter* there overnight, although he couldn't guarantee her safety as break-ins were by no means unknown. I decided to take my chance, so I pulled her on to the side of the slipway, covered her and padlocked her to a concrete post, and went in search of comfort.

The bed aspect was important, but beer was more so. I walked across the long, many-arched Victorian bridge that connects the two sides of town, and came to an irresistibly plain, ordinary, red-brick pub called the Bridge. One glance was enough. Inside, the barman, silver-haired and comfortably upholstered, advised me about beer, and I drank two pints of a pale, delectably zingy ale called Golden Delicious. They didn't do beds, so I recrossed the bridge to another pub that did, only to find that they had disposed of their last room while I was in the Bridge. The barman there directed me up the hill to a hotel. Two hundred yards, he said. It was a twenty minute trudge to the Edgecote, and extremely steep. But it was a reassuringly solid, ugly, comfortable place, with a re-assuringly genteel landlady, and they had a room and a bed with crisp, white sheets, and there was a big bath with old-fashioned taps and lots of hot water. I lay in the bath for some time and was happy.

Later, back at the Bridge, I resumed my conversation with the barman about beer in Burton. In the course of drinking my final two pints of Golden Delicious, I realised how lucky I had been – despite being in England's undisputed capital of brewing – to have stumbled upon this particular pub.

I doubt if even the most fervent Burtonian would argue that the town is a place of beauty. The newer parts are the usual straggling, characterless mess, while nineteenth-century Burton tends towards a monotony of straight, flat-fronted brick terraces. But in my view it is redeemed by the care it takes with its river.

On the map Burton resembles a butterfly, the two wings of housing, factories and commercial premises joined by a body of water, turf and woodland. Approaching the centre, the Trent splits into two channels enclosing meadows, with lawns sweeping down to the banks. There are footpaths and cycle paths, places to swim and row, saunter and snooze. Just below the main bridge, a lane cuts off past an old mill and some flats and houses, and leads down to a cricket ground and football and rugby pitches. Beyond is more shaggy meadowland, clasped between the two arms of the river, which are finally reunited beyond the boundary of the town.

On the evening I arrived, the floodplain seemed to breathe fresh oxygen into the lungs of the townspeople. These days the river's role is recreational and, arguably, spiritual. But in the past Burton's livelihood depended on it; while the nation – and places much further afield – depended on Burton to satisfy a simple, pressing need.

It began, according to the annals, with Frawin the Brewer, a tenant of Burton Abbey in the twelfth century.[6] He would have supplied the monks with their ale (made without hops, which were not introduced into the process until much later). By 1604 there were forty-six licensed alehouses in Burton and the reputation of Burton beer was spreading. It was available in the Peacock Inn, in Gray's Inn Lane, London, in 1670; and in a collection of popular songs printed in London in 1709 there is one that opens: 'Give us noble ale/of the right Burton pale/and let it be sparkling and clear.'

The water used to make the beer came, not from the Trent, but from wells to the north and west of the town. It percolated through layers of gypsum rock from which it derived high concentrations of magnesium and calcium sulphate. These minerals encouraged fermentation and the conversion of starch to sugar, enabling the brewers to add generous helpings of hops to create a clear, sharp-tasting beer.

However, while the river had nothing to do with the character of the noble ale, it had everything to do with getting the liquid where it was wanted. Burton's earliest export industry was not beer but alabaster, a pale and translucent form of gypsum much valued in the late Middle Ages for tombs, carved effigies and the like. It was transported in sizeable quantities down the Trent to Hull, from where it was shipped as far away as Italy.

Although Burton was classified as the highest navigable point on the river, at times of low flow, the shoals and shallows below the town made the passage of boats precarious, verging on the impossible. It was clear that if Burton was to take full advantage of its assets, the river must be made more reliable, and responsibility fell on the local bigwigs, the Pagets. In 1699 the 6th Lord Paget secured parliamentary approval for a scheme to dredge a channel up to Burton from Wilden Ferry, a few miles above Nottingham. He would fund the work in return for being allowed to charge three pence per ton for passage, and having the exclusive right to build and maintain wharves.

In the event, Paget was distracted by the demands on his diplomatic skills — he served as ambassador in Vienna and Constantinople — and he leased his rights to a pair of entrepreneurial spirits well practised in rough river work. One, Henry Hayne, owned the only wharf in Burton; the other, the much-abused Leonard Fosbrooke, operated the ferry at Wilden. Their business strategy was simple: river traders had to use their boats, and had to pay through the nose for the privilege. The situation was not accepted meekly, and there were endless disputes, protests, fights and dastardly deeds. But it prevailed until 1760, when the 8th Lord Paget — a man characterised by the *Dictionary of National Biography* as being 'chiefly remarkable for an inordinate love of money' — granted a new lease to the recently formed Burton Boat Company.

By 1789 Burton was sending 4000 tons of goods a year downriver. Much of this was beer, and much of that was destined for the Baltic countries and Russia, where they were particularly fond of a strong, dark, sweet ale made by Burton's leading brewers, the Wilsons. This Baltic trade was extremely lucrative, but also vulnerable. Napoleon's navy shut it off for a time, and although it revived briefly while the Emperor was imprisoned on St Helena, it was eventually finished off by the Russian government's decision to impose prohibitive import duties. Fortunately for Wilson's, in that same year – 1822 – one of their clever men managed to duplicate a process originated in London for making a much lighter, more bitter and sparkling beer with exceptional keeping qualities, specifically designed for export to India.

The superior quality of the Burton water, and the expertise of the Burton brewers – led by Samuel Allsopp, now in charge of Wilson's, and William Bass – enabled them to muscle in on the India Pale Ale trade. At the same time IPA became increasingly popular at home, and within a couple of decades the development of the national rail network had made it possible for Burton beer to reach almost every part of the land where there was a throat ready to receive it. By 1888 the Burton breweries were dispatching three million barrels a year, twice as much as London. Bass alone employed three thousand people and operated sixteen miles of private railway track within the town.[7]

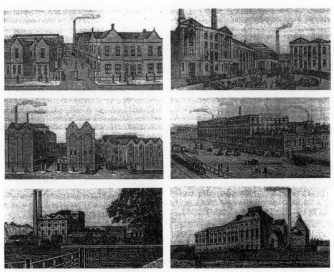

Burton breweries 1870s and 1880s from the top, clockwise: Truman, Hanbury, Buxton and Co, Black Eagle Brewery; Bass offices and brewery; Allsopp; Bass new brewery and maltings; Allsopp, Station St brewery; Mann, Crossman and Paulin, Shobnall Rd; Peter Walker, Clarence St.

The look of Burton was transformed. Ind Coope migrated from Essex, Charrington and Truman from London. New breweries, built in sturdy red brick and handsomely adorned with arcades, stucco gables and Italianate and pseudo-Tudor embellishments, sprouted in the centre. Vast malthouses as big as football grounds appeared on the outskirts where the barley was spread out to germinate.

The story since that golden – or amber – age, is a sad one: of mergers, closures, bankruptcies and takeovers; of the disappearance of comfortable, familiar names and the appearance of alien, ugly ones; of the abandonment or demolition of the handsome old brewery buildings and the rising from the rubble of gleaming and hideous new plants; most depressingly of all, of the coming of disgusting, chilled keg beers and acrid lagers, and the marginalisation of real, living ale. Burton's fall from grace was symbolised by the annexation of the Bass brewing business by the American company Coors. Thus did the Bass Museum of Brewing – with its old waggons and wooden casks and antique lorries built in the shape of a bottle of Bass bearing the famous red triangle trademark – become the Coors Museum of Brewing.

But there was light amid the gloom. Twenty-five years ago two managers at Ind Coope, fearful that real ale was about to follow the burbot into extinction, bought the defunct Fox and Goose beside the bridge and began brewing in the outbuildings at the back. It was into this sanctuary – which they properly renamed the Bridge – that I had strayed. Inside everything was plain and ordinary: wooden floor, wooden tables, wooden benches and chairs, wooden panelling, a burnished bar from which the pumps rose proudly. There was no music – 'this is not a fun pub,' the barman said sardonically – just the murmurings of chat from a disparate mix of young, middle-aged and elderly men and women united in being Burton people and enjoying what had been the pleasures of Burton people for centuries.

The barman and I agreed that there was much in a name. Allsopp and Martin and Bass and Worthington and Marston would all do very nicely for English beer-makers; but Coors? Never. I finished my pint and sauntered in the gloaming back over the long bridge. Immediately below it, an offshoot loops from the main Trent and tumbles over a weir before joining the secondary branch. The main flow itself continues some way downstream before crashing over a large and intimidating weir that I had prospected and didn't like the look of at all. But I thought the *Otter* and I might manage this more modest affair in the morning. I paused further on to check that she was snug and safe at the rowing club. Then I plodded up the hill to my hotel, concentrating entirely on the prospect of sheets and pillows.

Chapter 10

Crossing the Waters

An odd, fleeting confidence blossomed briefly at breakfast. The walls of the Edgecote's dining room were panelled in dark oak, around the top of which extended a display of plates and other items of decorative china. There was more crockery on display in a cabinet, beneath a spray of artificial flowers. Lace curtains restrained any view of the street beyond the windows. I felt inquisitive eyes on me as I sat down, and I turned to say good morning.

The one other guest was a woman, in her mid-fifties I guessed, with very carefully styled ash blonde hair, a long nose, delicate features. Our landlady came out of the kitchen, trailing fumes of frying and the strains of Radio Two. In a husky, refined voice, the other guest ordered a poached egg and a single slice of bacon. I felt slightly gross as I asked for an Edgecote traditional English in all its glistening glory.

While we waited we fell into conversation. I told her briefly what I was doing, and she registered a quizzical interest. She was from Malvern, and was in Burton on business in her capacity as a schools inspector. I learned that she was divorced from a French naval officer, 'a serial marrier,' she said with a disdainful little laugh. She had no children of her own, but was on good terms with his. She told me firmly that she was quite able to manage on her own. The landlady came in bearing plates, and heard this comment. 'Oh, I can't imagine not having a man about the place,' she said with a hearty laugh. The OFSTED inspector smiled dismissively and set about her poached egg.

People were streaming across the bridge on foot as I crept under it in the *Otter*. Several of them stopped to watch as I cut across to the weir. I grounded the punt, rolled up my trousers, and got out to manoeuvre her over the drop. There were a couple of dodgy moments as she swung in the current, one end stuck on the sill, the other swaying over empty air. But then she tipped forward and slid down, and I was able to wade with her and steady her before clambering in. I waved to the figures silhouetted against the sky above, and paddled off.

The stream flowed briskly over a wide bed of gravel thickly streaked with beds of water crowfoot. The sun was warm on my straw hat, and the water sparkled like pale ale. I had a fishing rod in the boat, and as I looked at it something stirred in the memory.

We had a favourite way, my brothers and I, of catching chub on summers' days on the river of our youth, the Loddon. You would kneel on the bank among the nettles and cowslips, swat the flies away, and swing out a ragged crust of bread so that it landed with a fat plop in a clear run between the tresses of weed. You would watch the crust intently as it bounced downstream on the ripples, paying out line as you did. Often nothing happened, and you would detach the hook from the sodden crust with a flick of the wrist, and start again. But sometimes there would be a swirl, and the bread would vanish into it, and a lift of the rod tip would be met with plunging resistance. Christ, it was exciting: you'd set the hook and leap up and charge towards where the chub was churning its way into the weed, yelling for someone to bring the net.

It occurred to me that this stretch of the Trent was just like the Loddon, but three times as big. There was another encouragement. Burton was full of swans – in fact they were promoted as a visitor attraction, and the civic authorities had even placed a huge and grotesque fibreglass model of one in a waterside park. Where there are swans you may be sure there will be people who think they are doing something for 'wildlife' by feeding them with bread. And where bread is thrown into the water, some of it will reach the stomachs of chub.

I had the remains of an elderly loaf. I broke off a couple of bits and tossed them towards the bank to see what would happen. A yard or so out from a trailing willow branch there was a loud suck, and the first crust vanished from sight. I stopped the *Otter* and tied her to a thick strand of weed, set up the rod, tied on a hook, and wriggled it into a chunk of bread. I swung it out towards the willow and let out the line. It arrived at the spot where the other crust had been taken. I was still, extremely tense, intent on the window-pane of water. The pane shattered, and the bread was drawn under. I struck, the rod bent like a bow, and I felt the steady, heavy pull of a good chub. After a couple of minutes I landed it, a fish of about three and a half pounds. I held it in the water for a moment or two, admiring the mail of silver scales along its flanks, its grey back and ivory-white stomach, its wide head and rubbery lips. Then I slipped it back, and watched as a flick of its tail propelled it to a safe place.

I looked around, hoping that someone might have witnessed the triumph. But there was not a soul about, so I just sat for a time and savoured it all on my own. I thought how lucky it was that I was here now rather than, say, 1855, when the condition of the river was Burton's disgrace. A letter of complaint to the Trent Fishery Association grumbled about 'hundreds of tons of stinking refuse' being emptied into the river, and of 'cartloads of fish' being killed. The Association reported that the condition of the water was so bad that at times angling had been impossible because of the quantity of waste, 'anglers being unable to sink their lines to the bottom . . . and even if this feat was accomplished, the bait would be covered in filth and no fish would take it'.

The river's condition was improved a few miles below Burton when it was joined by the Dove. It was entirely fortuitous that the Dove, even in the darkest days of Victorian river abuse, should have remained largely unpolluted; it just happened not to run near too many significant sources of toxic waste. The attitude of the locals towards conservation can be gleaned from an account of December 1853 of salmon making their way up the Dove to Tutbury in the hope of spawning, only to be shot at from the bridge 'by which they obtained several weighing twelve and fourteen pounds, while others were speared at the weir'.

'Was you ever in Dovedale?' Byron asked of his friend Tom Moore, adding: 'I assure you there are things in Derbyshire as noble as in Greece or Switzerland.'[1] It's true that Dovedale is some distance from Newton Solney, which is where the Dove joins the Trent and where I got my only glimpse of it. But I felt that the mingling of the waters brought the stories of two fishermen together as well; and that, having spoken of Walton, I should consider Charles Cotton too, particularly as it is easier to warm to him than to old Izaak, and the best of both was in their friendship.

The Cottons lived at Beresford Hall, on the Staffordshire side of the Dove a mile or two upstream of Dovedale proper. The old prints show it to have been an imposing mansion, but the fortune that paid for it was frittered away on a mania for point-less litigation which periodically gripped Cotton's cultured but fatuous father. By the time young Charles inherited, the estate was sinking under the weight of mortgages and other debts. The son was almost as improvident as the father; but he loved the place, not merely as a refuge from his creditors, and it grieved him profoundly to lose it – as he did – and to forsake the river:

Oh my beloved nymph, fair Dove,
Princess of rivers, how I love
     Upon thy flowery banks to lie,
And view thy silver stream,
When gilded by a summer's beam!
And in it all that wanton fry
Playing at liberty
     And with my angle, upon them
The all of treachery
I ever learned, industriously to try.[2]

I love that phrase 'the all of treachery': it suggests perfectly the combination of expertise and cunning that every successful fisherman needs. Cotton was a real angler, who learned as a boy to create whisps of feather and fur to trick the trout and grayling. His father was a contemporary and friend of Walton's, and the shared passion for running waters brought the old man and his young disciple together. Cotton was proud to address the monumentally virtuous Walton as 'father and friend'; perhaps the more so as he had a reputation for loose living (the mean-spirited Sir John Hawkins, Boswell's rival as Johnson's biographer, found his verses full of 'such foul imagery, such obscene allusions, such offensive descriptions, such odious comparisons, such coarse sentiments and such filthy expressions as could only proceed from a polluted imagination and tend to excite loathing and disgust').

Walton ignored the stories and stayed true. In accepting the twelve chapters on fly-fishing that Cotton contributed to the final version of *The Compleat Angler*, he declared his love and his intention to visit his friend 'though I be more than a hundred miles from you and in the 83rd year of my age'.

Cotton left a wonderful monument to their friendship. He wrote: 'I have lately built a little fishing-house upon it, dedicated to Anglers, over the door of which you will see the first two letters of my Father Walton's name and mine, twisted in cipher.' The house, with its steep, pyramidal roof, is still there, as is the cipher stone over the arched entrance, the initials entwined above the legend '*Piscatoribus Sacrum*'. A short cast away is the river itself. It is not at all difficult to picture the two anglers there, Cotton enjoying his favourite breakfast of a pipe and a glass of ale, listening respectfully to the white-haired old man, occasionally venturing a suggestion about the use of a Whirling Dun, dubbed with the bottom fur of a squirrel's tail and the grey wing feather from a drake; or a Green Drake in mayfly time, made from bear, sable and camel hair, hog's bristles and a mallard's wing feather dyed yellow.

Willington Power Station

Passing by the mouth of the Dove, the *Otter* and I left Staffordshire and entered Derbyshire. Downstream, two landmarks detached themselves from the flat countryside: the five brooding, silent cooling towers of Willington Power Station on one side; and on the other, the needle-slender spire of the Church of St Wystan, which pokes up above the ancient public school of Repton a mile to the south-east.

Willington was an important river port before Brindley's Trent and Mersey Canal bypassed it and put it out of business. Crockery from the Potteries was brought here in crates slung over packhorses to be loaded on to barges bound for Hull. Consignments of Cheshire cheese passed through, and salt from the Cheshire mines, iron goods from Staffordshire, hosiery and silk cloth from Derby. Coming back, the barges brought flints from Gravesend, plaster, wine, tobacco and choice groceries from the London retailers.

Twyford

My morning wore on. I floated peacefully through green fields and past innumerable grazing cows. The sun was hot, and the bubbling and burbling of the water along the *Otter*'s bottom was very soothing. A minute settlement appeared on the north bank clustered around a little church. It was called Twyford, the same name as the village in Berkshire where I grew up, so I decided to stop and have a look. There was a ramshackle farm, a terrace of cottages, the church and graveyard. Beside the river was a handsome cream-coloured Georgian residence with a notice at the end of the drive: DO NOT ENTER WITHOUT AN APPOINTMENT DOGS LOOSE, but there were no dogs or humans around.

Like Willington, Twyford was not always so inconsequential. A clue to its useful past was half hidden on the bank: a sturdy vertical post, buttressed from the sides. On the opposite bank I could make out the route of what had been, in another age, the packhorse road. The post now serves to hang a brace of faded lifebelts, but it used to support the great iron wheel that drove the Twyford chain ferry. Until 1963, when the ferry ran for the last time, this was a vital place for anyone who wanted to get from one side of the river to the other.

Men have aspired, where possible, to cross rivers over bridges. The building of a bridge both symbolised the power to overcome nature's anarchic tendencies and expressed civic spirit, the ambition to embrace the future. It acted as a kernel for settlement and growth. As well as enabling armies to move around, it offered connection, a clasp across the water, a way to bring people and places together to talk and do business. Their drawback was that they were laborious, time-consuming and extremely expensive; and in the back of everyone's mind was the awareness that the bridge was never as permanent or indestructible as it might seem

Destruction of Kelham Bridge

at the proud moment of its inauguration. Newark lost its first bridge in 1486 and several more subsequently. The arches of Nottingham's bridge were forever being damaged or washed away and repaired piecemeal. Kelham Bridge was demolished by an ice floe in 1855. Almost a hundred years later Cavendish Bridge at Shardlow collapsed in the floods that followed the great freeze of 1947.

Ferries had the advantage of being a lot cheaper than a bridge and slightly less hazardous than a ford. Otherwise there was not a lot to be said for them, and what was said was generally impolite. Daniel Defoe had to cross the Tamar estuary to get from Devon to Cornwall and counted himself 'well escaped' when he got to shore at Saltash. Faced with the prospect of crossing the Severn, Defoe's heart quailed: 'The sea was so broad, the fame of the Bore so formidable, the wind also made the water so rough that none of us cared to venture.'[3]

Thrumpton Ferry 1906

More than eighty historic crossing points have been identified on the Trent, two-thirds of them ferries and fords, the remainder bridges. In ancient days the lowest was at Littleborough, near Gainsborough, where a ford paved with slabs of stone took the

Roman road between Lincoln and Doncaster across the tidal river. Until Gainsborough acquired its bridge in 1791, the only transit for many miles was the ferry at Walkerith. The hazards of using it were demonstrated by a tragedy in 1761, when a traveller in a hurry found the ferry full and 'was so rash and imprudent as to leap his horse . . . and with the violence of the fall drove the poor people and their horses to the further side, which instantly carried the boat into the middle of the stream and overset it'. Six people were drowned.[4]

Radcliffe Ferry

With one notable exception, the ferrymen of history and literature tend to be an anonymous bunch. Travellers like Defoe tended not to ask for names and personal details, just to hope that their pilot was not drunk or suicidal, and that he knew the currents and the tides. The exception, of course, is mythology's most dismal figure, Charon, ferryman of the Styx.[5]

Working out the geography and organisation of the Greek realm of the dead is not easy, the confusion compounded by the habit of using the same name twice over. Thus Hades is both the god of the netherworld, the heartless rapist of Persephone, and the place itself. The Styx is one of the rivers of hell, but also its goddess and nymph, a daughter of Oceanus who lives in a house built on columns of silver overshadowed by soaring mountains. In the *Iliad*, Homer placed Hades in the far west, beyond the River Oceanus which encircles the earth. In subsequent versions it migrated underground, approachable via various chasms and separated from the realm of the living by its rivers – as neatly summed up by Milton in *Paradise Lost*:

> . . . four infernal rivers that disgorge
> Into the burning lake their baleful streams –
> Abhorred Styx, the flood of deadly hate;
> Sad Acheron of sorrow, black and deep;
> Cocytus named of lamentation loud
> Heard on the rueful stream; fierce Phlegethon
> Whose waves of torrent fire inflame with rage . . .[6]

As for the infernal boatman, he doesn't figure in Homer's cast at all, even though there is plenty of underworld action in both the *Odyssey* and the *Iliad*, in which the rivers act as barriers to be crossed before the wandering souls can find rest after death. Charon did not make his debut by name until he appeared fleetingly in a sixth-century B.C. Greek fragment called the *Minyas*,[7] and more substantially in plays by Euripides and Aristophanes. But it was Virgil, in the *Aeneid*, who turned him into the boating world's paramount anti-hero.[8]

Virgil played fast and loose with the mythological geography. He placed his underworld beneath Italy, and made significant adjustments to the riverine layout. Acheron, which appears to be both river and lake, is the principal barrier for the souls to cross – 'the river of Tartarus . . . a vast quagmire of boiling whirlpools which belches slime and sand into Cocytus, and these are the rivers and waters guarded by the terrible Charon'. As for the Styx, it is river and swamp and endowed with 'nine-fold windings' with which it somehow encompasses 'the waters of death'. In some dimly realised fashion, the infernal rivers all seem to flow in and out of each other, with Charon in overall charge.

He is not a pleasant sight – 'terrible in his squalor . . . on whose skin grows a thick grey beard never trimmed . . . a foul cloak hangs from his shoulder'. His eyes glitter and glare, and although he has the look of great age, his divine status guarantees inexhaustible strength to go with his immortality – 'with his own hands he plies the pole, tends the sail and in his ferruginous boat conveys the dead'. He is responsible both for transport and for selecting those who qualify for passage from the crowds stretching forth their hands from the shore. Aeneas asks how the choice is made. The Sibyl tells him that only the buried are taken quickly; the helpless souls of the unburied are left to wander for a hundred years.

Rivers and the waters of the underworld were important components of the belief systems of many ancient cultures. Gilgamesh rows across the waters of death in search of immortality. The Assyrian hero, Gisdubar, is ferried to the realm of shades to seek advice from his ancestors. Polynesian myths tell of a gulf that souls must cross either by canoe or swimming. The ghosts of the Karens of Burma must find their graves beyond rivers. Recent excavations near Stonehenge in Wiltshire have revealed that there was a substantial Neolithic village nearby enclosing a timber circle from which an avenue led to the waters of the River Avon. Archaeologists have theorised that the bodies of the dead were consigned to the waters, and that some were then removed for burial within Stonehenge itself.

In *Water and Dreams*, Gaston Bachelard suggests that 'water is needed for death to keep its meaning of a journey'; that 'death is a journey which never ends, an infinite perspective filled with dangers'.[9] Thus, in Bachelard's view, Charon's boat is 'a symbol that will remain attached to man's indestructible misfortune. It will pass through ages of suffering.'

* * *

Having floated downstream from Twyford and left the shadows of its departed ferrymen behind, I came to a cool, watery cavern beneath the arching branches of a great weeping willow where I lunched on a piccalilli sandwich, a slice of fruit cake and a glass of red wine. Afterwards I lay at one end of the *Otter* with a cushion under my head, looking up at the tracery of leaves and the blue sky beyond, listening to the tree and the rippling of the river.

Lunch place

Later, rowing along the southern edge of the village of Barrow-upon-Trent, I passed a man fishing from the bank. He was using an elderly cane rod and a reel of old-fashioned design known as a centrepin, which you don't often see these days. I asked him how he was getting on. 'Only just started'. 'Do you fish here much?' 'Five or six times a year.' 'What do you get?' 'Roach, perch, dace mostly. The odd chub. I'm after a barbel now.' Barbel are big, powerful river fish that take some catching. Nine out of ten Trent anglers I questioned were 'after barbel'. I was almost past him now, but I wanted to know about his reel. 'Had it since I were sixteen,' he said proudly. 'Should see you out then,' I said. 'And

Thirsty work

me grandchildren.' I wished him luck. When I looked back he was sitting still on his stool, one hand holding his rod, the other feeling for a pluck at the line.

I ended my day beside Swarkestone Bridge, a lovely and unusual structure condemned to a daily living death by the unbroken stream of traffic that roars over it from dawn to dusk on its way between Derby to the north and the M42 motorway to the south. There is a smart pub next to the bridge called the Crewe and Harpur, after the ancient Derbyshire family – for a time just Harpur, later Harpur-Crewe – which used to rule the roost around here. The pub had a lawn sloping down to the river, and the sight of it rekindled my enthusiasm for beds, baths and the like. I concealed the *Otter* in the bankside vegetation, padlocked her to a tree, then emerged on to the lawn rather like one of those Japanese soldiers who used to pop out of the jungle every few years, explaining that no one had told them the war was over.

The resident Harpur at the time of the Civil War was Sir John. Before leaving to fight for his king, he stationed a garrison to defend the bridge. They failed, and the Parliamentarians proceeded to wreck Sir John's seat, Swarkestone Hall. In more peaceful times he had commissioned an eccentric building with two castellated towers and a ground-floor arcade to look out over his bowling green. Known variously as the Pavilion, the Stand and the Balcony Field, it is still there beside its walled rectangle of turf, looking as quaint as ever. On my way back from inspecting it I met an elderly man with no shirt on who was washing his car, his straggly white chest hair glinting with droplets of water and dabbed with puffs of soap bubbles. He told me proudly that Sir John's pleasure house had featured on the front of a Rolling Stones album, and that the group had actually performed there.

Swarkestone bridge long ago

Legend has it, in that improbable way that legends have, that the first Swarkestone Bridge was put up by two sisters as a memorial to their lovers, both of whom had drowned trying to cross the Trent in a flood to press their suits. There was certainly a stone bridge here early in the thirteenth century which had a little chapel on it. A church document of 1552 alludes to it suffering a deplorable fate '. . . whiche had certayns stuffe belonging to it, ii desks to knell in, a table of wode and certayne barrel of yron and glasse in the wyndos which Mr Edward Beaumont hathe taken awaye to his owne use and we saye that if the chappell dekeye the brydge wyll not stande'.

The chapel did indeed decay and eventually disappeared altogether. As for the bridge, a new five-arch structure was put in place at the end of the eighteenth century to span the water, but it was fitted into the medieval causeway that winds for fully three-quarters of a mile across the floodplain from the village of Stanton on the south side. I waited until after supper to walk along it. By then the traffic had dwindled enough for the bridge's charm to be appreciated. Beneath the winding walls of the causeway there were mysterious pools, the water black and still and thick with fibrous weed, the surface stamped with bright green lily pads.

I had a final pint in the pub garden, as the shadow of the bridge dissolved into the twilight. A cairn on the grass bears an inscription recording that this was the most southerly point reached by Bonnie Prince Charlie's army in 1745. In fact the bridge was held by an advance guard while the Young Pretender remained in Derby, arguing the toss with his chieftains as to whether they should go on or go back. The decision to retreat is usually attributed by historians to the failure of the French to provide the support they had promised, and to the reluctance of the English to rise up against the Hanoverians. But it's also been suggested that the Scots, far from their mountain strongholds, were demoralised by the sight of flat, featureless, colourless, drab Derbyshire. The poet U. A. Fanthorpe has a marvellous line: 'He turned back here. Anyone would.'[10]

Chapter 11

The Jolly Miller

Swarkestone Bridge from below

To my angler's eye, the water at Swarkestone was extremely tempting. It shallowed and quickened as it approached the bridge, below which it was broken by gravel beds into several bubbling, weedy, fishy streams. The main weight of the current swung into the south bank in an arc furled with eddies and miniature whirlpools, highly suggestive of chub and barbel. On my evening walk I had followed a car track down that bank and come upon a shaven-headed angler in waders, up to his thighs in the water. He'd had a couple of chub and had hopes of a barbel as the light faded. He told me of a previous capture in the same spot, a chub of seven pounds, which is very big indeed.

Swarkestone was cherished for its fishing long ago. One of the lesser poetic lights of the age of Donne and Marvell, Thomas Bancroft − known as 'the small poet' both because of his lack of inches and the scale of his verses − fondly remembered the

> Sweet river, in whose flowery margin laid,
> I with the slippery fish have often played
> At fast and loose.

Bancroft also dragged the river into a laboured classical allusion for the solemn epitaph he composed on the death of one of the Harpurs:

> As did cold Hebrus with deep groans
> The Thracian harper once lament,
> So art thou with incessant moans
> Bewayled by thy doleful Trent
> While the astonisht Bridge doth show
> (like an Arch-mourner) heaviest woe.

The nineteenth-century philosopher Herbert Spencer remembered staying with his parents in Derby, and being unable to sleep: 'I got up, dressed, sallied out, walked to Swarkestone five miles and began fishing by moonlight.' Spencer is not much considered these days, but in his day was a great friend of, and influence on, the likes of John Stuart Mill and Thomas Huxley. He was an early evolutionist and a ferocious libertarian, arguing among other things that if a citizen preferred to do without the state's provision of education, poor relief, ever water, he should not have to pay taxes. Spencer's radicalism endeared him to

Marian Evans — otherwise known as George Eliot — and their friends thought they might marry, but he was apparently deterred by her conspicuous plainness. His background was modest and he earned his living for some years as a railway engineer before turning to philosophy. Famously industrious, he suffered some kind of nervous collapse in his thirties and thereafter dictated much of his writing while rowing or between strenuous games of rackets. Fishing was another solace for his restless mind, although he has nothing interesting to say about it in his long and tedious autobiography.

I had a notion to emulate the poet and the philosopher. But to have any hope of success in that steamy weather, I had to be fishing by moonlight, or dawn at the latest; and by the time I came to in my soft bed it was half past six and the cars were already pouring in unbroken lines over the bridge. So I had my breakfast and then eased the *Otter* out into the current, leaving Swarkestone and its bridge to their unquiet fate.

It was warm and getting hot. The rhythm of the river took hold of us again, hurrying us over the broad, weedy shallows, slowing us over inscrutable pools. Across the fields the canal plodded doggedly north-east, while the course of the river twisted this way and that, as if asserting its right to do what it wanted. There was just one slight shadow over my contentment as we progressed, caused by the appearance of the word 'weir' on the map, and a reference on a canoeing website I had come across to 'King's Mills — first-class rapids'.

On the southern side the flat fields gave way to rising ground, thickly wooded. Squeezed against the slope, the river narrowed and quickened. Along the north bank, smothered in a thicket of willows, alders and brambles, I glimpsed stone walls, the remains of landing stages, a stone bridge over a sidestream. Ahead rose a pair of toothy, rusting wheels, their motion frozen. I held the *Otter* in mid-current. As usual, the crisis had galloped up on me without

Old mill wheels

giving time to prospect or plan. All I could do was follow the river's command and hope for the best.

Walls appeared, pinching the flow into a crooked bottleneck. We turned abruptly to the left, away from the rusty wheels, and I paddled like a demon to keep the boat straight and away from the slabs of crumbling stone. Then we shot through the gap and out into a wide pool. The main current curved towards the north bank, which is Derbyshire, while an eddy circled back to the south

Quickening

side, where for a few miles the Trent forms the western boundary of Leicestershire. I paddled into the slack water, thrust the *Otter*'s front end into a little bay and sat still for a few moments, waiting for my heartbeat to slow.

The setting was picturesque, in a manicured sort of way. Smooth lawns spread down to the river from a hotel that clearly boasted pretensions several cuts above those of the Edgecote. Ranks of limousines gleamed in the gravelled car park. Liveried flunkeys hastened to and fro. Mowers buzzed over the grass. Apart from the decayed wharves I had whizzed past, and the great wheels rearing silently at the water's edge, there were no obvious clues to the long and busy history of King's Mills.

Because of the Trent's susceptibility to flooding, most of its mills were on tributaries or artificial cuts. This was one of the few exceptions. The Domesday Book records that corn was being ground here more than a thousand years ago. In the sixteenth century it was a fulling mill, where woollen cloth was scoured to remove oils and shrunk and pounded to make the fibres more dense. Two hundred years later it was doing duty as a paper mill and had spread to both sides of the river. In the course of the nineteenth century it was adapted again, to grind gypsum from Chellaston, near Derby, into plaster, which was then barged downstream to Nottingham and points east. By then, though, it was feeling the pinch. Steam engines had left waterwheels behind, and the railways had elbowed the canals aside. King's Mills staggered on into the twentieth century, and its ferry service survived until the Second World War. But one by one the businesses gave up the ghost. The millstreams silted up, the wharves were swallowed by the march of vegetation, the wheels slowed to a halt, left as mute witnesses to the time when water was a power in the land, and the mill and the miller were as vital to the community as the church and the parson.

* * *

The process of grinding cereal grains to make them edible goes back ten thousand years. There are carvings on Egyptian tombs showing women turning an upper stone against a stationary concave saddle quern. The same technique, using local sarsen stone, was used in Wiltshire four millennia ago to grind emmer and einkorn, two primitive cereals. Iron Age engineers moved the technology on by adding a handle to the revolving stone and enclosing the gear inside a structure shaped like a beehive, with a hole at the top for the grain to be fed into.

The drawback was that the turning of the stones required human muscle power, a limited and valuable resource. Civilisation was crying out for a labour-saving device to go with the wheel and the *shadduf*, the counter-balanced water scoop. It is likely that the watermill was invented in Alexandria, the ancient world's principal centre of learning, research and inquiry. It was first described in the second century BC by Philo of Byzantium,[1] and his account was recycled in the tenth volume of *De Architectura* by that same ingenious Vitruvius Pollio who had observed that springs were fed by rainfall. The principle (discovered by Archimedes) was the use of a vertical wheel to turn a horizontal millstone by means of a connecting spindle. The technology spread throughout the Roman Empire, and eventually reached Britain, although the extent of its use here is a matter of debate since almost no Roman mills survived the upheavals that followed the collapse of Roman rule. If it did disappear, it was not for long, for there is a record in the tenth-century chronicles of Abingdon Abbey that the saintly Abbot Aethelwold — chiefly notable for having introduced strict Benedictine rules on monkish behaviour to England — commissioned a watermill to serve the monastery and a leat off the Thames to turn it.

Karl Marx identified the watermill as a crucial contributor to the development of Europe's feudal system, and the statistics make a compelling case. Domesday lists more than 5600 of them in

England, a number that had at least doubled by 1300. The mill became essential to the functioning of the community, and a source of political control, while the miller himself emerged as a formidable figure on the social landscape.

Thanks partly to Chaucer, the miller acquired a reputation for dishonesty and bullying that was probably no more deserved than today's caricature of parking wardens as power-crazed tyrants.[2] Millers were middlemen, beholden to the boss – the lord of the manor, generally out to maximise his income – but dependent for trade on the freemen and villeins who needed to have their corn ground. Some were employees of the manor, but by Chaucer's time most were tenants, paying the lord a fixed annual sum and levying a toll of between one-thirteenth and one-thirty-second on the corn milled. There were certainly dishonest millers, but it seems unlikely that, as a class, they were bigger crooks than reeves, summoners, pardoners or franklins – or, for that matter, parsons and monks.

The grinding power of the mill wheels could be readily put to other uses. Fulling mills appeared towards the end of the twelfth century, and mills were adapted to crush bark for dye, to make paper, grind gunpowder and flints, operate bellows in blast furnaces, prepare leather, tobacco and snuff, and saw wood. When the time came, it was around a watermill that the prototype of the industrial factory was built, and it was water power that drove the machines; all a few miles up the next Trent tributary below King's Mills, the Derwent.

The pioneers were Thomas and John Lombe, the sons of a Norfolk worsted weaver. John, who was the younger, was sent to Piedmont in northern Italy to learn his trade and surpassed expectations by coming back with a blueprint of the method developed there for throwing silk. He and his brother installed a giant waterwheel twenty-three feet in diameter on an island in the middle of Derby, and put up a five-storey building next to it to contain what their patent of 1719 described as 'a new invention of three sorts of engine never before made or used in Great Britain, one to wind the finest raw silk, another to spin, and another to twist'.

A few years after the silk mill started operating, John Lombe died in mysterious circumstances. According to the Derby historian William Hutton – who was apprenticed to the mill at the age of seven – he was the victim of a revenge plot hatched by Piedmontese silk-makers embittered at having their secrets stolen by the Englishman. Hutton claimed that they sent a seductive female agent to Derby, who wormed her way into John Lombe's affections and administered a slow poison from which, after suffering appalling agonies, he died. The story is probably fiction; what is sure is that the surviving brother became extremely rich, had a knighthood conferred on him, and was very proud of his mill.

It was famous. Boswell visited it in 1777 (although Johnson, who accompanied him to Derby, preferred to stay indoors with their host talking medical matters). Another visitor was Richard Arkwright – that 'plain, almost gross, bag-cheeked, pot-bellied Lancashire man', in Carlyle's words – who was looking for a power source cheaper than horses to drive his new spinning frame. Arkwright was impressed, and in 1771 he began producing woven stockings at his own water-powered mill a little way up the Derwent, in the steep wooded valley of Cromford. Two years later Arkwright graduated to the manufacture of cotton cloth, and the British textiles industry was born.

Water power inspired the revolution, but its time was short-lived. Arkwright himself installed one of Watts and Boulton's steam engines at his new mill in Nottingham in 1790. Before long the waterwheel as an industrial power source was following the canals into general redundancy. Even so, it continued to be the preferred method for grinding corn to meet local needs, and as the urbanised, industrialised society grew, so did the rural mill become a symbol – no longer of feudal oppression – but of a disappearing way of life. Poets and painters waxed sentimental about them. Constable fixed the image with his paintings of his father's mills at Flatford and Dedham. Tennyson remembered

> . . . the brimming wave that swam
> Thro' quiet meadows round the mill,
> The sleepy pool above the dam,
> The pool beneath it never still,
> The meal-sacks on the whiten'd floor,
> The dark round of the dripping wheel . . .[3]

The reputation of the miller himself soared. The Irish songwriter Isaac Bickerstaffe celebrated a figure Chaucer would not have recognised:

> There was a jolly miller once
> Lived on the River Dee;
> He worked and sang from morn till night;
> No lark more blithe than he.

Instead of being associated with crookery and a tendency to swagger and bully, he came to epitomise dependability and joviality. Close to the water, close to the earth, the miller spoke his mind and showed a distrust of modern ways as deeply ingrained as the flour that whitened his meaty hands – qualities all embodied in the doomed figure of Mr Tulliver in George Eliot's *The Mill on the Floss*.

Something of the old, quieter England was lost when the waterwheels stopped turning. Edward Thomas sighed sadly over

Water that toils no more
Dangles white locks
And, falling, mocks
The music of the mill-wheel's busy roar.[4]

Of the two Thames mills nearest my home, one is a theatre, the other – almost miraculously – is still grinding corn. It is at Mapledurham, and its appearance would have stirred Constable: a tumbledown, timber and brick building, a high, dripping wheel, and – reflected in its pool – the neighbouring Jacobean mansion. Twice a week the mill is charged into life. The great stone turns and the flour pours down its chute as it has for a thousand years and more. It is excellent flour and it makes excellent bread. But there's no pretending that it is needed, or that it plays a central role in a community's existence. It is someone's hobby, that's all – pretty, quaint, irrelevant.

There is a powerful passage in John Stewart Collis's *The Moving Waters* that mourns the loss:

All has perished under the assaults of energy. And the whole manner of living that went with it has been swept away – all that wonderful family continuity and community which was its essence. I do not say it was better than the present manner; for, though I may think so, how can I be sure? There was plenty to weep and wail about then, as now. The lords of the manor knew how to grind the faces of the poor as well as the corn. Better or not, a whole form of living has gone – on balance, quite as good as ours, we may be sure. They knew peace. They heard music in the almost organic hum from the turning of the great wheels which were all they had of machinery and factories. If we wander about in the countryside we can find them still, many of them – hidden in dark, woody corners, at the ends of sequestered lanes, by the sides of abandoned weirs. There is peace here, but it is the wrong kind of peace; in that silence there is a terrible unease. The door is open, for there is no one to close it. The windows are broken. The great stones lie forsaken and the paddles stuck in weedy bonds. Outside, upon the water, a lonely swan, bereaved and mateless, moves into the shadow of a tree bent low with splintered age. The flowers do grow, the birds do sing; but no human sound is here. The present is ousted by the past; but the place is a haunted vacancy; it is without life; it is null; it is nothing save a voiceless lamentation for the ruin and the rout.

I sat in the *Otter* for some time pondering the shades of history while the water rushed past. There was more history up the hill behind me: another sad tale, I suppose, but one worth the trouble of following.

The census of 1841 reveals that seventy members of the parish of Castle Donington – which includes King's Mills – were employed as boatmen or watermen. Many more worked at the mills themselves. Castle Donington itself, a couple of miles away to the east, was a prosperous market town, feeding off the river and canal trade and its proximity to the main roads between Leicester and Derby, and between Birmingham and Nottingham. Close to the town was the grand mansion, Donington Hall, seat of the Hastings, who had had their hands on the levers of influence around here for many centuries.

The Hall was built by the 1st Marquess of Hastings in the 1790s on the proceeds of successful campaigning in India. Although a grateful East India Company had voted him £60,000, the sum proved inadequate to fund the house and his other extravagances, and the estate was already heavily mortgaged by the time it passed to his son Francis, the 2nd Marquess.

Fox-hunting was his passion in life, but that life didn't last long, and his heir's was shorter still, so that by 1853 the marquisate of Hastings, established only twenty-seven years before, was already on to its fourth – and, as it turned out, final – holder.

The coming of age of Henry Weysford Charles Plantagenet Rawdon-Hastings in July 1863 was marked with great celebration in Castle Donington and at the Hall. There was a fireworks display, a dinner of roast beef and plum pudding, and an entertainment for two thousand schoolchildren in the park. In December the Marquess opened his own racecourse, and staged the Castle Donington steeplechases.

He was soon making a stir beyond sleepy Leicestershire as well.[5] On 17 July 1864, *The Times* reported the news that had already electrified fashionable London. Lady Florence Paget, star of society and fiancée of a rich and handsome Lincolnshire landowner, Henry Chaplin, had eloped with the Marquess of Hastings and been installed as mistress of Donington Hall. The details were exquisite. The day after accompanying Chaplin to the opera, Lady Florence entered Marshall and Snelgrove's store in Oxford Street on the pretext of buying items for her wedding trousseau, sneaked out of the back and into Hastings' carriage and was whisked off to church where the couple exchanged their vows. The marriage, reported *The Times*, 'was a hurried and unexpected one – more particularly, it would appear, to the connections of her Ladyship, none of whom were witnesses to the ceremony'.

By then Hastings was already well down the road to ruin. Although he had been temporarily distracted by a two-legged filly, his deepest passion in life was horses: not riding them (an overprotective mother would not let him) but betting on them. To the bookies, the tipsters, the trainers and the host of rackety chancers drawn to the Turf, he was Harry Hastings the Plunger. To add to the gambling disease, he developed a pathological obsession with getting the better of the man whose girl he had stolen, Henry Chaplin.

The rivalry came to a sensational head on Epsom Downs in June 1867. The day was bitterly cold, and as the thirty runners for the Derby were led into the paddock, heavy snow was falling. Chaplin's horse, Hermit, had proven class but was prone to burst blood vessels, and had drifted to 66 to 1. The Plunger had plunged: £100,000 with the bookmakers on his colt, Uncas, plus a personal bet of £20,000 with Chaplin. A succession of false starts delayed the off for an hour, adding to the tension. Finally the field broke. Hermit's jockey, John Daley, followed his orders to the letter, bringing him through in a long, late run to win by a neck.

Hastings was finished, financially and socially. The bookies rubbed their hands and dispatched the duns to Leicestershire. The Marquess's aristocratic friends turned their backs on him. There is a story that Lord George Bentinck, the senior steward of British horse racing, came to dinner at Donington Hall and found himself seated next to a notorious bookmaker, named Padwick. Pressed for an explanation by an outraged Lord George, Hastings said it was 'such a convenience' to have him there. 'So is a night stool,' snapped Lord George. 'But one doesn't have one in one's dining room.'

A year after Hermit's Derby victory, Hastings was booed and hissed at Epsom following the breakdown of his overraced filly, Lady Elizabeth. By then his own health had also collapsed, and he died the following November at the age of twenty-six. His obituary in *The Field* was not charitable:

There was something approaching to insanity in the way in which he scattered his means. He had not even a sportsman's excuse for his prodigality. He had no personal prowess, was no horseman, and cared little or nothing for the hounds he kept for a season or two . . . whilst the thousands he wagered on a plating race might, so far as real sport was concerned, as well have depended on a straw or the colour of a card.

The crash was sorely felt in this corner of Leicestershire.[6] Although Hastings' brother-in-law managed to hang on to the house, the rest of the estate had to be broken up to meet the debts. The library of twenty thousand books, the wine, the plate, the furniture and more than five hundred paintings were sold, as were the various businesses, including the factories at King's Mills. The whole structure of patronage and employment was destroyed. As one observer put it sadly: 'Racing may be the sport of kings but it is a sorry amusement on which to fling away a fine estate like Donington Park.'

Donington Hall

These days Donington Park is best known as a famous venue for open-air rock and heavy metal concerts. The track created by Hastings for his own steeplechases is a motor-racing circuit, and the air is split by the snarl of engines. On a pole above the grand entrance to the mansion of the Marquesses of Hastings flutters the flag of the budget airline BMI. It is a parable of our times.

Chapter 12

'I constrained the mighty river'

Peace below King's Mills

I was regretfully aware, as we drifted down over the shallows below King's Mills, that the nature of the journey was about to change. Thus far my progress down the river had been pretty much carefree. There had been the occasional alarm, such as at Hoo Mill, but nothing serious, and we had been good companions, the river and I. We had kept together from the soggy meadows below Biddulph Moor, through Stoke, through the dairy fields of Staffordshire, past churches and cooling towers, beneath bridges and over shoals and deep, silent pools; and all the way I had had the feeling of being partners on an adventure, myself as the discoverer.

Apart from the kayakist at Stone and the sculler at Burton, I had met no one on the water and precious few beside it. The river had been unlike any road or path. On it, I had been absorbed into the landscape. I had felt and heard the life of the water as I followed its surprises. By confining me within its banks it had freed me from my attachment to the unmoving earth and made me a different spirit. And all this way it had managed to hold on to its own spontaneity, the sense of not being answerable to any master, wandering as it would. That was all coming to an end.

The first warning was an insistent buzzing ahead. A low, flat, ugly bridge appeared with two dark eyes for arches. Its name, the Cavendish Bridge, is an insult to its predecessor, a fine, five-arched affair built in the 1760s at the behest of the Duke of Devonshire and named after his family, which lasted until the 1947 flood did for it. The tablet displaying the old toll charges has been retained: two-and-sixpence for coaches, chariots, landaus etc with four wheels, sixpence for horses, mules and asses 'not drawing', a penny for pedestrians unless they were soldiers, in which case they paid a halfpenny.

The building of the Duke's bridge reflected changing times. Until it opened, those taking the turnpike road had to get out of the carriages and down from their horses and surrender themselves to the mercies of the Fosbrookes, who operated the Wilden Ferry. Like the Haynes upstream, the Fosbrookes were not fastidious when it came to protecting their business; when a consortium of Nottingham traders challenged his exclusive right to provide passage, Leonard Fosbrooke put a boom across the river and summoned forty thugs to man it.

But the days of freebooting and lawlessness were drawing to a close. The vast manufacturing potential of Midland England was beginning to be tapped. Great fortunes were being made and others were there for the taking. Investment was pouring in, and the returns depended on transport. Small-time freelances like Leonard Fosbrooke and Henry Hayne belonged to another era and had to be discarded. The wealth of the nation was at stake.

In September 1772 James Brindley lay dying at his home in Turnhurst, north of Stoke. A mile or two away, the navvies were still digging their way through Harecastle Hill to bring the northern section of the Trent–Mersey Canal to his devoted friend Wedgwood's factory at Etruria. But the 'Schemer' had lived long enough to see much of his vision realised, and to get an idea of the way it would change the face of England's heartland.

Rather than cut the canal through to Burton, as the brewers had begged him, Brindley had decided that it should join the Trent further east, at the junction with the Derwent. Downstream from there, the river was – or could be made – reliably navigable. A place called Shardlow, hitherto a small and sleepy farming community, was chosen to be the hub of this waterwheel.

The Derwent was crucial. Apart from adding 40 per cent to the volume of the Trent, it was already an important trade route in its own right. As early as 1719, the year the Lombe brothers patented their silk-throwing machine, Parliament had approved a bill to make the Derwent navigable from Derby down, to assist 'traders and dealers in lead, butter, cheese, malt, marble, grindstones, scythe-stones, iron, timber and other commodities by reason of being a much cheaper and easier conveyance than at present by land carriage'.

Two years later the first barge arrived in the town carrying a cargo of timber, tobacco and fish. Derby went wild with excitement. The band played and the town newspaper reported that there were 'Ringings of Bells and Demonstrations of Joy . . . it must be acknowledged that there were never any Rejoycings in which the Inhabitants have been so unanimous . . .'

Brindley's decision to make Shardlow his canal terminus turned it into a boom town. To reach the distant markets now made accessible, goods had to be shifted off the big, broad Trent barges on to the narrowboats able to negotiate the canal locks, and vice versa. Warehouses sprang up. Basins and inlets were dug. Wharves were built and cranes and hoists were installed. Wealth suddenly poured into Shardlow. A visitor in 1789 wrote: 'Around the navigation are built so many merchants houses wharfs etc sprinkled with gardens looking to the Trent and to Castle Donington as to form as happy a scene of business and pleasure as can be surveyed.'[1]

But Shardlow had its uncouth side. A Moravian pastor, Karl Moritz, walked through much of England in the course of 1782 and drew a vivid picture of the habits, conversation and behaviour of the country folk and working people he met. The Navigation Inn at Shardlow made a deep impression on him: 'A rougher or ruder kind of people I never saw. Their language, their dress, their manners were, all of them, singularly vulgar and disagreeable; and their expressions still more so. For they hardly spoke a word without adding a G— or a D— to it, and thus cursing, quarrelling, drinking, singing and fighting, they seemed to be pleased and to enjoy the evening.'[2]

Thanks largely to the Schemer, the early enthusiasm for canals blazed into a passion. Within a few years of the opening of the Trent and Mersey, a new canal, the Erewash, was ready for use. This entered the Trent a couple of miles below Shardlow, at Sawley, and gave access to the Nottinghamshire coalfield and the fuel to power the Industrial Revolution; which was excellent news for the Willoughbys of Wollaton, who sat on the best of it and had been growing richer and richer on coal for three hundred years.

The first of them, back in the thirteenth century, was actually a Bugge, but he did not care for the name so he appropriated one of a more genteel character. One notable Willoughby was Sir Hugh, who sailed in search of the North-West Passage to India and got as far as Finland, where he and his crew starved to death in the winter of 1553–4, the baronet's body being found some years later by Russian fishermen, seated in his cabin with his will spread out on the table in front of him. Another was Sir Hugh's great-nephew, Sir Francis, who inherited the estates and mines at the age of thirteen, and when he came of age commissioned one of the grandest Elizabethan mansions in the land, Wollaton Hall, said to have cost the colossal sum of £80,000.

Wollaton Hall 1880

The Willoughbys combined aristocratic distinction with commercial nous. They had their own fleet of barges to ferry Wollaton coal down to Gainsborough and Hull, and by 1600 had installed a primitive horse-powered railway to get the coal from the mines down to the river. When the proposal for a canal was first mooted, they were instrumental in pushing it forward.

The canals opened the way into the heart of manufacturing England and fired a wealth-making explosion.[3] Brindley's successor as the country's foremost water engineer, William Jessop, surveyed the Trent from Shardlow down to Gainsborough and set about ensuring a dependable channel. Lead, copper, ironware, cheese, salt, beer and pottery were shipped in ever-increasing quantities down to Hull at the mouth of the Humber. Upriver came Swedish iron, oak from Norway, flints from Gravesend for the china-makers, malt, hemp, flax, wine and groceries from the London merchants. But coal was by far the most important freight. By 1816 one and a half million tons a year was being transported to Hull from the coalfields of Nottinghamshire and Derbyshire, now linked via an extension of the Erewash Canal to Cromford.

It's estimated that 140 barges regularly plied the Trent. Most were around seventy feet long and capable of carrying forty tons of goods. They were generally pulled by horses, although they carried a single, square sail to use on the wider stretches when the wind was favourable. Hull, Gainsborough, Newark and Nottingham prospered on the trade, as did many smaller settlements – none more so than Shardlow. One old-timer recalled his days as an apprentice there:

> All the day through might be heard the creaking of cranes, the rattling of chains, the falling of timbers, the shouts of the boatmen and wharf-men ... There were the sounds of the hammer, axe and saw, the sound of the anvil, and the well-known noise of boat-builders ... The whole night through, at short intervals, might be heard the rattling of the coach wheels on the road and the merry notes from the bugle horn of the stage-coach or the red coat of the Mail; on the canal the same, for the fly-boats of Pickford kept on in a like manner through the night as well as day, and many of those seemingly illiterate men might be heard discoursing the sweetest of music which, in the night-time, was delightful to hear.

The music didn't last long. On 16 August 1832, the Nottinghamshire coal-owners met at the Sun Inn in Eastwood and resolved to build a railway line across the Trent to Leicester to move the coal that had hitherto taken the more leisurely route down the Erewash Canal then up the River Soar. This was the embryo of what, within a few years, became the Midland Railway, a network linking Nottingham, Derby and Leicester with London. As its tentacles spread, the canal system withered. In Shardlow, the warehouses shut down, the wharves crumbled, the basins were filled in, the boat-builders went elsewhere or gave up, and the licensing trade went into irreversible decline.

But at least the Navigation Inn, where Pastor Moritz was so roughly received, does survive. It's a handsome, whitewashed building which stands beside the old turnpike road. I was ready for the worst when I went in, but it was empty apart from a pasty-faced barman deeply engaged with the *Star*. I had a pint of Marston's, thought about lunch, then thought again. Canned pop music washed through the empty spaces that had once resounded to raucous balladeering, tankards being slammed on oak tables and the shouting of unspeakable oaths. Something had definitely been lost.

Shardlow is now the Shardlow Wharf Conservation Area. The Salt Warehouse demurely retails heritage. Next door, the Clock Warehouse has been converted into a bar/restaurant; the space once occupied by the clock is an empty eye socket, and the arched loading bay underneath is a toothless mouth. A boating trade of a genteel kind is carried on, servicing the narrowboat holiday sector. The quays and towpath are lined with *Lady Janes*, *Leyandras*, *Celestines* and *Lady Isabellas*, the sun flashing across the polished livery of bottle green, royal blue and burgundy, names picked out in gold lettering, every brass fitting agleam, terracotta pots and hanging baskets awash with reds and purples. The din and the drunkenness, the fighting and whoring and swearing that so offended Pastor Moritz have given way to murmured exchanges concerning the weather and the opening times of the locks, and the muted chug of diesel engines.

Along the towpath are spaced the iron mileposts, scrupulously painted in black and white at the instigation of the heritage people. On the last one are the names that shaped James Brindley's vision – Shardlow and Preston Brook – and the distance between them, ninety-two miles. It's a long way to dig.

In his analysis of the greatness of cities, Giovanni Botero declared that God had created the waters 'for a most readie . . . means to conduct and bring goods from one countrie to another'. The concept of the divinely appointed trade route was attractive but flawed. In fact, to obtain the kind of rivers Padre Botero thought most desirable – those 'calm and still', on which boats might 'saile up and downe with incredible facilitie' – required human intervention on a grand and sustained scale. From Botero's time onward, the hydrological engineer played a critical role in shaping the manageable, productive landscape demanded by progressive societies.

Across Europe, major rivers were shortened, dredged, embanked, channelised and generally brought to order. The Trent Navigation Company's chief engineer, William Jessop, organised cuts to avoid notorious shoals, as well as the building of new locks and weirs and an intensive programme of dredging designed to ensure that the barges could get through even in low water. The Trent – in common with many European rivers – was forced to assume a new character. It gave up its spontaneous, unruly side. Little by little, the navigable reaches became uniformly broad, flat, opaque and ponderous. Little by little its capacity to surprise and make trouble was taken from it (although it is only in the last fifty years or so that this process of subjugation has been completed).

That is how it is now, bar the occasional flight of fancy, downstream from the junction with Brindley's canal and the Derwent. As it passes under the M1 motorway bridge, which squats over it like a brown toad a couple of hundred yards below Shardlow, the river drops below the 100-foot contour. For the rest of its journey to the

Humber, it belongs to its wide, flat valley. Hemmed in between its flood banks, it rolls quietly and respectfully through towns and cities and villages and farms. In its well-behaved way, it is impressive in the way large rivers are impressive, but to those who have known it in its careless youth also rather bland, pompous, slightly dull.

Inevitably officialdom marches hand in hand with responsibility. The Trent was now a Navigation, which meant I needed a licence for the *Otter*. To get it, I stopped at Sawley Marina, where there is a British Waterways office empowered to issue such documents. The marina is situated halfway along an artificial channel known as Sawley Cut, which was dug across the bottom of a troublesome northern loop in the river's course. I tied up to a bollard near the petrol pumps and scrambled up a ladder on to the concrete quay, leaving the *Otter* bobbing gallantly between two floating pleasure palaces. Stretching away before me was an enormous fleet of water craft: barges, yachts, catamarans, cruisers, dinghies, powerboats. But no punts.

The British Waterways office was at one end of a store selling ropes, rowlocks, life jackets, maps, distress flares and everything else essential for a day, or a life, afloat. I filled in a form given to me by a friendly woman, who gave it to a less friendly woman, who said I needed to show proof that the *Otter* was insured. Various sun-tanned boating types wearing canvas shoes and sharply creased slacks and Cowes-style caps drifted in and out with items of chandlery while we wrangled. In the end I had to make a call to British Waterways' headquarters to establish that, since the *Otter* didn't have an engine, she didn't need to be insured. Having got the licence, I left and was accosted by a man who leaped out of a battered van and asked me in a strong Eastern European accent if I wanted to buy a generator. I made rowing motions at him and said I didn't need electricity, but I don't think he understood.

Seized by an anxiety to escape from bureaucrats, salesmen and the leisure and recreation industry, I jumped in the *Otter* and began rowing hard. I had to be careful, though, as there was a steady stream of narrowboats ploughing their way upriver, and I had my back to them. I tried to keep close to the bank, but then I had to be wary of the anglers. They were out in force, one every 15 yards or so, like sentinels watching out for a rumoured invasion. Every now and then one would bend to pick up his rod, reel in and cast again. No one seemed to be catching anything, but the conditions – blazing sun, oven-like heat – were terrible. I passed one old chap dressed only in a flattened sun hat and a pair of Union Jack boxers, from which his legs descended like sticks of dead wood.

Cooling

The Erewash Canal joins the Trent at an uninteresting place called Trentlock, opposite the mouth of the Soar. The sky, blue everywhere else, was white with the steam from the Ratcliffe Power Station, although the six cooling towers were obscured from my lowly line of vision by a lump of rising ground known as Red Hill. Here I had to follow another boring artificial cut, the alternative being to negotiate a large and alarming cataract at Thrumpton Weir, which my friend at the Environment Agency had urged me not to try. As a result I rejoined the river close to an ancient and historic mansion, Thrumpton Hall – but not close enough for me actually to see it. Hot, bothered and weary, I forgot about it altogether, until it was too late.

THRUMPTON HALL,
The Seat of John Wescomb Emmerton Esq.

Its great treasure, by reputation, is a grand and intricately carved staircase of oak, elm and pine that rises from the basement up to the attic. It was commissioned in Cromwell's time by the owner of Thrumpton, Gervase Pigot, who also played a small but vital part in launching an enterprise for which all students of the history of Nottinghamshire have reason to be grateful.

ROBERT THOROTON, M.D.

It was while staying with 'our good friend Mr Gervas Pigot', wrote Robert Thoroton in the dedicatory epistle to his *The Antiquities of Nottinghamshire*, 'that I was persuaded to write the History of this county'. Thoroton was a delightful, early specimen of that admirable breed, the local historian. Born at Car Colston, a few miles from the southern bank of the Trent between Nottingham and Newark, he became a doctor. But finding himself unable, as he put it, 'to keep people alive for any time . . . I decided to practise upon the dead', which was a quaint way of describing his mission to record every fact he could unearth about the lives of the departed great and not-so-great, and good and not-so-good. It was a monumental task, which he embarked upon in earnest in 1667 and finished ten years later, a year before his own death.

I would have liked to have paid my respects to the steadfastly Royalist Thoroton and his Puritan friend, and to have had a look at Thrumpton's red-brick gabled front. But by the time I remembered about them I was well downstream and not inclined to row back. By then the current was no longer giving me any perceptible help at all. I pulled and kept on pulling, mental activity dulled by the rhythm of the dipping oars, the creaking of the rowlocks, the stretching of muscles and sinews in my arms and shoulders. In front of me the sun was sinking towards Derby, and somewhere over to my right was Attenborough, where Cromwell's right-hand man, Henry Ireton, was born. The village is on the edge of a vast, reedy, bird-infested network of old gravel workings, along the western side of which creeps the grubby and insignificant River Erewash. A few miles up it is the mining village of Eastwood, where D. H. Lawrence was born and spent his generally miserable childhood.

Lawrence studied geography at Nottingham University College, which presumably included something about rivers, estuaries, floodplains and the like, but he doesn't seem to have had any particular enthusiasm for the subject. Both his local stream, the Erewash, and the Trent itself were horribly and dangerously polluted at the time, so perhaps he can be excused his lack of interest in them. There are a couple of Trentside scenes in *Sons and Lovers*.[4] In one, the deeply sensitive and unlovable Paul Morel takes a walk upstream from Wilford with his lover, Clara Dawes, on a dark, rainswept afternoon. Beside them 'the river slid by in a body, utterly silent and swift, intertwining among itself like some subtle, complex creature'. They take the path through Clifton Grove, then slip and slide together down the steep slope to 'the full, soft-sliding Trent' where Morel makes love to her, impelled by an emotion that 'was strong enough to carry with it everything – reason, blood, soul – in a great sweep, like the Trent carries bodily its own back-swirls and intertwinings, noiselessly . . . everything borne along in one flood'.

The Lawrentian flood tide of passion seemed more than usually unreal on a sun-soaked June evening, the river sunk in reflective lethargy, lazily winding between the gravel pits on one side and the flat meadows of its old floodplain on the other. In the middle of the meadows, at a safe distance from the water, is the dead-end, unbothered village of Barton-in-Fabis, whose curious name is a Latinised version of Barton-in-the-beans, referring to what was once the chief crop of the fertile fields around.

In medieval times, Barton was the property of the Greys of Codnor, which is a few miles north-west of Eastwood. The last of them had no legitimate heir but a tidy number of bastards, among them two boys – 'the greater Harry' and 'the lesser Harry' – one of whom got Barton. The property later passed to the Sacheverells of Derbyshire, and it was William Sacheverell who led local opposition to James II's plan to impose a new charter on Nottingham. He was hauled up before the King's favourite enforcer, Judge Jeffreys, to answer for his defiance. Jeffreys addressed him thus: 'You should know better than to ask such insignificant, impertinent questions as you do. It was very saucy, I tell you . . . we are come to a fine pass that every little prick-eared fellow must come to demand maces that are the badges of authority, and they must be told, forsooth, that they are saucy.'[5] The prick-eared Sacheverell was fined 500 marks for his sauciness; but he had the satisfaction of outliving both the bully Jeffreys – who died of alcoholic excess in 1689 – and the monarch who had sent him.

Cabins at Barton

Until the 1960s there was a ferry service at Barton, catering mainly for day trippers from Nottingham who wanted to saunter along the river and up through the bean fields to take refreshment in the village's tea shops. Where the ferry used to call, there is a line of chalets and cabins which looked very cheerful and sweet as I rowed past. Each had its little pontoon and most a boat of some kind. Some owners were tending to their craft, one or two were fishing, rods angled against a darkening sky, others were just sitting out on their patches of grass, glasses in hand.

A little way below this tranquil community of river people, the Trent divided around a thickly wooded island. The official navigation was to the west, and the nearside channel was narrow, silted and undisturbed. I paddled the *Otter* about halfway along it until

I found a little bay where I could make camp. It was a delightful spot. Behind me arable fields extended towards a ridge carrying the main road to Clifton and Nottingham. In front, the trees on the island — dominated by a cluster of soaring white willows — screened the boat procession along the main channel. I put up my tent, poured the wine and heated a pan of baked beans and spam. I was settling down to enjoy the peace and the feast when a speedboat approached from downstream and released on to the island a gang of noisy and high-spirited teenagers. Over the next couple of hours they became noisier and more high-spirited as their store of empty beer cans grew. Most of the time they were content with pushing each other into the muddy shallows and wallowing riotously around. At one point they tried to bring their boat up the channel towards me, but they ran aground and had to turn back.

It was almost completely dark by the time they flopped merrily back into their boat and departed towards Nottingham. I listened to the sound of the engine as it faded into the night. The air was utterly still, the black outlines of the trees against the charcoal sky as unmoving as if they had been sculptures. The river was too slack to make any noise of its own, and the coots and moorhens were quietening down. Peace settled once more over the village in the beans and the fields where the beans once grew.

Chapter 13

Poets and Cannibals

The next morning I listened, as I did every morning, to the weather forecast on Radio Four. The forecaster, Jay Wynn, reported that storms were marching up the country from the south-west. Some of the rain, he cautioned with relish, would be torrential. Where I was, the sky was a soft, innocuous grey, the progress of the river smooth and serene. Nothing seemed to threaten, but I thought I could sense an instability in the air around me as the pressure fell. Maybe if I hadn't heard the forecast I wouldn't have noticed anything.

Only the waterfowl were out and about as we glided past the tail of the island and rejoined the main river. Around the next bend, at Beeston, I had to make a choice. The official navigation, the sanctioned route for responsible craft, veers away to the north through a lock and along an artificial channel known as the Beeston Cut. After passing a housing estate and the kingdom of Nottingham's most famous company, Boots Pharmaceuticals, it goes under the A52 and joins the lower section of the Nottingham Canal through the southern fringes of the city before abruptly turning back to the river and rejoining it just below Trent Bridge.

My alternative was to stick with the river on its old, unruly path to the east. Immediately opposite Beeston lock, the Trent tumbles over a concrete weir shaped like a clumsily bent bracket into a big, foamy pool. At the tail of this pool it collects itself and hastens over a wide gravel bed to push against the foot of the steep, tree-blanketed escarpment of Clifton Grove where its energy soon diminishes as it returns to a north-easterly course.

I didn't hesitate for long. I had no particular desire to linger in Nottingham, which had long ago decided – for its own sound reasons – to keep its distance from the river. The founders had spread it across two hills a mile away to the north, with its castle on top of one and its parish church on the other, leaving the broad, flat space across to the Trent as a buffer zone and sponge to soak up the winter floods.

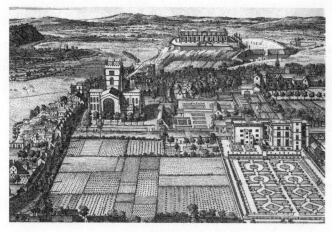

Nottingham 1709

Believe it or not, Nottingham was once famed for its position and looks. Defoe called it 'one of the pleasantest and most beautiful towns in England', and the doughty Pastor Moritz enthused over its 'lofty houses, red roofs and glittering spires'. The pastor took a 'charming footpath' across the meadows to the river, eventually reaching an inn where the company was conspicuously more friendly and decently behaved than in Shardlow, even though all he could get by way of dinner was bread and butter. That was in the 1780s, and for a time Nottingham people continued to have good reason to be proud of the beauties of their city and its splendid Market Place, and the orchards and rose gardens spread over the approaches to the river, and the riots of blue crocuses that presaged every spring. One remembered 'the ivied lanes, the scented fields, the misted glens of long ago', wandering through the meadows between the Trent and its tributary, the Leen, 'watching the fish in the clear waters and the dragonflies skimming along the surface, or gathering flowers that grew along the banks'.[1]

Nottingham 1750

Such innocence could not last. The rapid expansion of Nottingham's world-famous lace industry caused its population to soar, and sent factories and terraces of cheap cottages creeping out from the old Lace Market towards the river. The Leen met the usual fate — one local poet wrote of it in 1847 as 'dark, pestilent and clouded/All noses that now pass thee by/In handkerchiefs are shrouded'. Five years later, Mrs Ann Gilbert, the doyenne of Nottingham's literary community, composed a lament called 'The Last Dying Speech of the Crocuses' whose concluding lines recorded the arrival of 'The Spirit of Trade':

Come line and rule — come board and brick — all dismal things in one/Dread spirit of Inclosure come — thy wretched will be done.

Done it was. The railway arrived, and across the fields spread a thick rash of sidings, sheds, more factories, more terraces of mean cottages. The area kept its old name, the Meadows, but it became synonymous with disease, squalor, crime and misery: the worst of slums. It was also, as it always had been, vulnerable to the river. A modest flood was enough to inundate the swarming tenements, sending untreated sewage swilling through the streets. When the great freeze of January and February 1947 gave way to the great flood of March 1947, five thousand homes and commercial premises in the Meadows were marooned. The city council ordered new flood defences to be built, and took the opportunity to sweep away the slums. In their wisdom, they then decreed that the whole area should be built over again. So the name lives on, now a byword for gangs and guns and drugs rather than crocuses.

No: I had no strong urge to engage with the city if I could avoid it. I inspected the weir. The water made a good deal of noise as it crashed over the drop, but along the side there was no more than a trickle. I pulled the *Otter* around the edge of the boom above the weir, unloaded her, and guided her over the sill and down the steps to an eddy at the bottom. I put everything back in, and we edged out into the current and I felt again the energy and spirit of a river temporarily released from its restraints. I had been rowing for a long time and I knew there was a lot more rowing to come. But now, for a little while, I did not have to row. I could sit, facing downstream, using the paddle to keep us away from the weedbeds and waterlogged branches and tree trunks, letting the river do the work. I kept one eye on the water, another on the almost sheer side of Clifton Grove as it loomed above us.

The Grove gets its name from the family that owned pretty much everything around here for eight centuries or so. The first was Sir Alvered Clifton, a fine, knightly name for one of the Conqueror's band; and one of the last was Sir Juckes, which also has a touch of the Round Table about it. In between, the senior Clifton was generally either Sir Gervase or Sir Robert, and an assortment of each rests in the village church.

Like the Bagots of Blithfield, the Cliftons of Clifton were

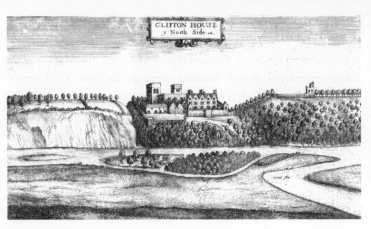

Clifton Hall 1676

generally content to attend to their lands and their people and not to seek distinction on the national stage. But kings and queens always knew they could count on them, in matters trivial and serious. In 1634 Charles I wrote to Sir Gervase Clifton asking him to use his good offices on behalf of his favourite, the poet Sir John Suckling. Sir John wished to marry the daughter of Sir Henry Willoughby of Wollaton. Anne Willoughby, however, did not wish to marry Sir John, who was very good looking – John Aubrey particularly admired his beard which 'turned up naturally so that he had a brisk and graceful look' – but feckless and thoroughly unreliable. She turned to another of her suitors, Sir John Digby, who challenged his rival by the road and 'cudgelled him into a handful, he [Suckling] never having drawn his sword'.

The last of the Cliftons was the most colourful.[2] As a young man, Robert – the pampered only son of Sir Juckes – shot, chased girls, drank, lived the high life and bought, sold and gambled on horses in reckless style. He spent some years in France and Turkey, dodging his creditors, and it took the death of his father to bring him back to Nottingham at the age of twenty-six. Once there, he conceived a passion for politics and put himself forward for Parliament, pledging to fight plans to restrict the hours alehouses could stay open. Sir Robert stated his case simply: 'I respect temperance. I respect good order. I am not a teetotaller. When I am thirsty I drink, and I hope every man in this room when he is thirsty or hungry will have the best of everything to eat and drink.' His eloquence appealed strongly to Nottingham's drinkers and he was returned with a huge majority. When he died of typhoid in 1869, aged forty-two, unmarried, heirless, up to his neck in debt, twenty thousand Nottingham people lined the way from Clifton Hall to the church to say farewell.

The Hall is a plain, handsome, unpretentious English gentleman's residence, which was not looking its best when I dropped by as it was in the process of being turned into a hotel. It stands a little way down from the village of Clifton, which has managed to hold on to many of its old buildings and some of its old charm, despite being squeezed between Nottingham Trent University and the Clifton housing estate. However, the glory of Clifton was, and still is, The Grove.

This great swathe of broad-leaved woodland was originally planted around three hundred years ago along the crest of the escarpment that rises like a surfacing whale from the Trent's eastern bank. It survives much as it was made, with paths and bridleways wandering back and forth between the oaks, beeches, limes and sycamores, with only the odd burned car wreck as a reminder of the city beyond. Saplings have spread down the precipitous slope to the water's edge, mixing with brambles, nettles, thistles and other humbler forms of vegetation. From the top, glimpses of the sparkling, speeding river are rare. Even so, it is easy to appreciate how some romantic soul of poetical disposition might be moved to sit here and ponder the scene, and – stirred by the Muse – dig into his pocket for notebook and pencil.

The Trent's overall record as an inspirer of verse is not impressive. The Dove had Cotton, the Severn had Housman. Coleridge was spurred by the Otter (and the Alph), Wordsworth by the Duddon, the Derwent and the Wye. Hardy studied the Dorset Stour, noting 'the reticulations . . . that creep on the slack stream's face'. Burns wandered the banks of the Nith — 'the winding stream I love so dear' — and Scott versified tales of Tweed and Yarrow. Rupert Brooke dipped his finger in the Granta, and Cowper loved the slow, winding Ouse.

As for the Thames, it has stimulated a genre all of its own. For Elizabethan and Jacobean poets, eager to make sense of England's past, glorify its present and ingratiate themselves with royal and noble patrons on the way, the premier river was charged with unique geographical and symbolic importance. They used it as a route into history and myth, and as an allegory for what everyone hoped would be new-found prosperity, splendour and peace. The centrepiece of the fourth book of Edmund Spenser's *Faerie Queene* is the wedding of the Thames and the Medway. The ceremony is attended not merely by the four rivers of Eden, but the Nile, the Scamander, the Tigris, the Euphrates, the Orinoco, and every English and Irish river of any note whatever (including 'the bounteous Trent', 'that in himselfe enseames/Both thirty sorts of fish and thirty sundry streames') — all of which rejoice to acknowledge the Thames as 'their Principall'.

Spenser's contemporary William Camden inserted into his *Britannia* a poem celebrating the wedding of Thames and Isis, the mingled flow representing male and female unity, father and mother in one. Less well known, but equally typical, was the *Degli Eroici Furori*, written in 1585 by the unfortunate Giordano Bruno, who was to end his days roasting over a bonfire in the Campo dei Fiori in Rome on the orders of the Inquisition. Bruno lived in England for a few years and produced this allegorical fiction about nine young men in search of perfect beauty as a not-very-oblique hymn of praise to Elizabeth I. In it, the Thames is portrayed as the heavenly river and its nymphs are endowed with an impressive array of magical powers.

Against all this (and much more) the Trent can offer Shakespeare's lines about being 'smug and silver'; a passage from Drayton's topographical epic *Poly-Olbion* expanding on Spenser's passing reference ('A more than usual power did in that name consist/Which thirty doth import; by which she thus divin'd/There should be found in her, of Fishes thirty kind;/And thirty Abbeys great . . . and thirty several streams' etc); and Henry Kirke White.

Trent bard

Who? I hear you say. White was the son of a Nottingham butcher who took up the pen in preference to the cleaver, and died of consumption at twenty-one. Byron thought highly of him, highly enough to remember

> Unhappy White! While life was in its spring,
> And thy young muse just waved her joyous wing,
> The spoiler swept that soaring lyre away
> Which else had sounded an immortal lay.[3]

Bard's home

Byron considered him as good as Chatterton, although reading White's best-known poem now, you wonder if he might have been thinking of someone else. This was '*Clifton Grove*', in which the poet looked down upon

> . . . the deepening glen, the alley green,
> The silver stream, with sedgy tufts between,
> The massy rock, the wood encompass'd leas,
> The broom-clad islands and the nodding trees.

Bard's cottage again

There is a good deal of similar stuff, woven around the tragic tale of the Fair Maid of Clifton, the beautiful Margaret, who betrayed her lover while he was off fighting in France, driving him to drown himself in the Trent when he got back, whereupon a demon dragged her to her doom in the same 'mutual grave'.

A darkening sky chased me past those wood-encumbered leas from which Fair Margaret plunged. Thunder muttered as the unappetising dun slab of Clifton Bridge came into view. I was under it as the first flurries of rain came, and under it I stayed as the storm raged. As I looked upstream the way I had just come, a prodigious fork of lightning split the bulging sky. The wind blew up then down in furious gusts, the rain hissed into the surface of the water, the thunder boomed. It was a splendid piece of meteorological theatre that I would not have enjoyed half as much had I been exposed to it. After half an hour or so, the tumult ceased and I cautiously resumed my progress.

The next bridge down is a slender affair installed by Sir Robert Clifton to connect the village of Wilford with the city of Nottingham. Nearby is a very dark and gloomy church, with a gloomy graveyard. A window in the church commemorates the

Capt and Mrs Deane at rest

sickly and neurotic Kirke White, who had asked to be 'sepulchred' here, where '. . . the Zephyr/Comes wafting gently o'er the rippling Trent/And plays about my wan cheek'. Among those at rest outside, in a tomb surrounded by a rusty railing, are Captain and Mrs Deane, who died within twenty-four hours of each other in August 1761. They were both at an advanced age, although in the case of the captain, there were times when he must have doubted if he would ever see his native land again, let alone find his peace in the churchyard of the village where he was born and brought up.

The tablet on the tomb tells the more respectable part of his story, recording that between 1714 and 1720, he 'commanded a ship of war in the Czar of Muscovy's service, after which, being appointed by his Brittanick Majesty Consul for the Ports of Flanders and Ostend, he resided there many years and by his Majesty's leave retired here to this Village'. That official career was adventurous enough – at one point Deane was imprisoned in Siberia for losing two of the Czar's ships, and much of his time as consul was spent spying on alleged Jacobite conspirators. I can picture him puffing on his pipe in the Ferry Inn, Wilford, regaling the regulars with tales of daring and intrigue in places most of them would hardly have heard of. But I wonder if he was as forthcoming about an earlier episode in his colourful past.

In 1710 Captain Deane set sail from London for Boston to trade a cargo of cordage. His partner in the venture was his brother Jasper, and they called their ship the *Nottingham Galley*. On 11 December she was wrecked off the New England coast, and the surviving crew took refuge on a bare lump of rock called Boon Island, near the mouth of the Picataqua River. They had no fuel, no protection from the elements other than a torn section of sail, and nothing to eat, a want they addressed by resorting to cannibalism. After almost a month they were rescued, and soon afterwards sensational stories were circulating about the cause of the wreck and the events on Boon Island. The *Nottingham Galley*'s mate published a version in which he accused Captain Deane of every imaginable brutality and of deliberately driving the ship on to the rocks to claim the insurance.[4] Other accounts lingered with relish over the scenes of cannibalism, referring to the dead carpenter being 'used up'.

Deane and his brother responded with their own *Narrative of the Sufferings etc* in which they vigorously repudiated all charges. They subsequently fell out with each other, and for several years carried on a vicious fraternal feud that ended only when Jasper Deane ruptured a blood vessel while attacking his brother, and fell dead to the ground.

WILFORD BRIDGE & CLIFTON STATUE, NOTTINGHAM.

Wilford bridge and Sir Robert Clifton

Sir Robert Clifton's toll bridge is at the top of a big loop in the rippling Trent's course. The river then turns south, and the area enclosed between it and The Meadows is a public open space partly occupied by war memorial gardens, which are separated by a carriageway from an expanse of turf dotted with trees leading down to a stepped embankment made of brown concrete. The council is evidently very proud of this amenity, which is certainly popular with runners, cyclists and people looking for somewhere to have a sandwich or for their dog to defecate. From river level, though, its effect is depressing. The perspective is dominated by the embankment, which is plastered with dark excrement from the gangs of swans and Canada geese that honk and hiss and flap all over the place, without anyone apparently giving a thought to controlling their rampant population growth.

Trent at Wilford 1914

The embankment was still wet from the last downpour when I passed it. Another was clearly on the way, but it held off long enough for me to leave behind the dispiriting bulk of County Hall and the flattened arches of Trent Bridge. Rowing diligently and keeping close watch on the state of the sky, I thought about my one previous visit to Nottingham, on a roasting day in that roasting summer of 1976. I was twenty-five then and full of hopes and ambitions never realised and now forgotten. I must have walked over the river from the station to the Trent Bridge cricket ground, where England were playing the West Indies. Did I see Vivian Richards flaying his way to 232? Or did I see the England rearguard led by a silver-haired, bespectacled plodder called David Steele, who saved the day with a six-hour century? I have no idea; all I can remember now is the baking, crushing heat.

I couldn't see the cricket ground from the *Otter*. But I did get a clear view of another gathering place for sporting shadows, the City Ground, its Brian Clough Stand a gaunt reminder of the long-departed era when Nottingham Forest were a force in the land.

Suspension bridge

Downstream from the greasy gates of Meadow Lane lock, where the Nottingham Canal creeps in, the last of Nottingham's river crossings appeared, Lady Bay Bridge. As I approached it, the rain came. I paddled over to the north side and tethered the *Otter*. Upstream and downstream the rain came down in sheets, but in the shadows it was dry. I ate yet another piccalilli sandwich and yet more cake, poured myself a glass of wine, and settled down to read *The Mill on the Floss*.

I was deep in the tale of the Tullivers of Dorlecote Mill when I became aware of company. A gangly youth was looking down at me from the bank. I said hello and explained what I was doing there. He said he was from Mansfield, and that he came down on Fridays because there was an activities centre. I asked what kind of activities. 'Kayaking and stuff,' he mumbled. A man in shorts and canvas shoes bustled down carrying rope ladders. He introduced himself as the leader of the Arches Adventure Base, a project funded by the council and local businesses to give youngsters something healthy and harmless to do. He didn't specify what kind of youngsters, but I gathered they might be the troublesome kind. One of the business sponsors turned up and helped attach clips to the underside of the bridge and unroll the ladders. He said the activities were geared to problem-solving and team-building.

I asked them to keep an eye on the boat while I slipped around the corner to a petrol station shop to get a paper and a few basics. By the time I got back there was a knot of youths and girls standing around under the bridge, waiting for something to happen. Some of them were wearing safety helmets and harnesses and looking slightly embarrassed. There was also a black man who was older, old enough for his hair to be tinged with silver. The project leader told everyone to get ready. It was a team-building exercise – they'd be climbing rope ladders blindfolded, relying on others to guide them. The black man hung back. He told me he'd tried it the week before and didn't like it. His name was Daniel.

He came and stood over me as I sat in my chair. He stared at the boat and asked what I was doing. When I told him, he laughed incredulously. I had to repeat myself several times before he would accept that I was travelling down the river alone, in this.

'You mean you're on your own, all day, all the time?' he asked, shaking his head. I said there was a lot to be said for solitude. For one thing, you didn't waste time and energy on discussion, disagreement and recrimination. You only had yourself to blame if something went wrong. 'I couldn't do it, be on my own like that,' he replied. He asked me what I did about cleaning my teeth and taking a shower, and I said I managed the first but stayed pretty dirty. Daniel found this hilarious.

'So you writin' a book?'

'That's the idea.'

'What kind of book?'

I explained that I wanted to write about rivers – not just this river, but rivers in general. He clearly thought I was taking the piss. I told him I might put him in it. He laughed delightedly. He asked me what I thought about when I was on the water. The past, I said. The mistakes I'd made.

'Do you repent?'

I felt suddenly awkward. 'I haven't done anything that bad.'

Daniel looked at me sceptically.

Behind him the ladder-climbing session was winding up amid laughter and good-humoured mockery. There was a nice atmosphere around the little group. The man from the sponsoring company, pink in the face, told me how important it was to show them that there could be more to their lives. I wanted to hug him for his generosity of spirit.

The rain had stopped and the cloud was lifting, although streams of water were tumbling from the bridge supports. They went off to fetch the kayaks, and I got ready to be on my way. Daniel lingered to see me off, and we shook hands.

'I think we meet again,' he said as the gap between us widened.

' I don't think so, Daniel. I shouldn't think I'll be coming this way again.'

He smiled. 'I mean in heaven.'

Chapter 14

First Love

Castle burns

It was a night of black storm, orange flames and destruction. The tenth of October 1831. Those who witnessed it never forgot it.

The spark was the arrival from London of the Pickford's van. It brought the news that the House of Lords, bullied by the Duke of Wellington, had thrown out Lord John Russell's Reform Bill, which would have extended the franchise and abolished some of the smallest and rottenest of the boroughs. In Nottingham the mob, habitually drunk, frequently disorderly and a constant source of fear and trembling to all respectable citizens, was hungry for trouble.

The following morning, Sunday, an excited crowd gathered outside the White Lion to greet the mail van from the capital. One of the passengers shouted that the reformers were beating to arms. A cry went up and missiles were flung at the houses of Nottingham worthies known to oppose the widening of the franchise. The Mayor, Mr Wilson, hurried from church to make an appeal for calm, and was attacked and beaten. Magistrates ordered that the Riot Act be read, and for a time the rabble dispersed. But they were not in the mood to miss their fun.

As darkness fell the next evening, Monday the 10th, disturbances broke out. A troop of the 15th Hussars was summoned to clear the streets. One band of rioters swarmed off to wreck a flour mill. Another marched east, intent on showing one of the city's most prominent anti-reformers, Mr John Musters, what they thought of him and his opinions. As it happened, Mr Musters was not at Colwick Hall, his splendid mansion beside the Trent. But Mrs Musters was, with her son, one of her daughters, and a companion, a Swiss lady, Mademoiselle de Fay.

It was a wild night, in every way. Rain was beating down as the invaders burst through the gates and gathered outside the north front of the house – 'uttering wild, appalling shouts . . . their bad passions inflamed by intoxication', as one of the many lurid accounts of the riot put it. They ripped up the railings around the lawn, smashed the windows, then burst in. While Mrs Musters and the others hid in the ballroom, the mob set to work. The fine furniture was smashed, vases, plates and a precious collection of Chinese porcelain were reduced to smithereens, a painting of a nymph by Titian was slashed, a Rubens was mutilated and two Canalettos were burned. Gunpowder was detonated in Mr Musters' dressing room, and family papers were looted or set on fire. The damage was put at £3000, and about the only crumb of comfort was that the portraits by Reynolds of Mr Musters' father and mother were spared.

At some point the ladies escaped through the ballroom windows into the shrubbery nearby. From there, cowering unseen, they watched through the hissing rain as the orgy of destruction proceeded. At length, 'the enfuriated desperadoes receded . . . their appalling shouts rolling their awful burthen on the breeze'. They had had their amusement at Colwick Hall, and news had reached them that Nottingham's most conspicuous symbol of the power and privilege of the ruling class was now under siege.

The Castle had looked down on the city for nearly eight hundred years. Built on the orders of William the Conqueror, it had been useful to various kings over the ages. John had twenty-eight boys hanged from the battlements because their fathers, Welsh chieftains and nobles, had risen in revolt. Richard III marched out from its gates to meet his fate at Bosworth Field. Charles I raised his standard there, not that it did him any good.

After the Restoration, Charles II gave it to one of those who had supported his father through thick and thin, the Duke of Newcastle. The Duke — Europe's leading expert on dressage and author of the standard *La Méthode et Invention Nouvelle de Dresser les Chevaux* — set about turning the battered shell into an enormously grand mansion. It was eventually completed after his death, but the Newcastles never lived there, and by the time of the riots the building was in a state of advanced delapidation. Nevertheless, it remained the property of the family, then headed by the 4th Duke, whose views on political reform and other matters were so extreme that he made the Duke of Wellington seem like a radical.

In driving rain, the crowds stormed the Castle. Fires were started in several places, and by seven o'clock the building was alight. This was the abiding nightmare of the decent, law-abiding classes: the rabble on the rampage. Children were held up to windows to witness it. One recalled the spectacle:

Beneath, a dense black mass of human beings, more like wild beasts, shrieking and howling: above, columns of smoke through which the flames penetrated, lighting up St James's Church, Standard Hill, and the whole neighbourhood. Athwart that deafening roar came the sound of crackling and falling timbers. One impressive sight was the molten lead pouring down in lurid streams from the roof of that noble mansion.[1]

Nottingham's spasm of anarchy was short-lived. The mayor ordered a curfew. Soldiers and special constables patrolled the streets in force, and the mob dispersed. The initiative was now back with the authorities and they acted in the usual way. Three of the supposed ringleaders were hanged before a large crowd in the centre of the city. When one of them, a local prize-fighter called George Hearson, stepped up to the gallows, he took an orange from his pocket and threw it towards St Mary's Church. As it fell, a small boy ran forward to catch it. Moments later Hearson's body was swinging at the end of a rope.

After the ransacking of Colwick Hall, Mrs Musters was escorted by servants to the stables, where she spent the night in one of the grooms' beds. The next morning she left for another of her husband's stately homes, Wiverton Hall, which was far enough away to the east to be safe from marauders. She died there two months later at the age of forty-seven, and it was said that she never recovered from the shock she sustained that rain-swept evening. It was a sad end to a rather sad life, one that would have been very different – and quite possibly even sadder – had she listened to flowery protestations of love from another quarter before Mr Musters led her up the aisle.

Mary Chaworth was, at seventeen, the loveliest as well as the richest heiress in Nottinghamshire; and George Gordon Byron, the 6th Lord, aged fifteen, with a deformed foot, no money and a fury of a mother, loved her with an intensity that made him ill.

Byron's love

Much later, Byron was to claim that he had overheard her say scornfully to her maid: 'Do you imagine that I could ever care for that lame boy?' Given Byron's lifelong enthusiasm for the theatre of cruelty — where beautiful, heartless women take cold delight in trampling on the poet's tender heart — it seems quite likely that he invented the scene, with the dual purpose of illustrating female callousness, and explaining why Mary Chaworth should have preferred Mr Musters to him.

He was everything Byron was not: rich, tall, athletic, bodily complete. He was a superlative horseman, and believed unquestioningly that, for one fortunate enough to be born a country gentleman, the pursuit of foxes should come before any other duty or obligation. 'Nimrod', the great chronicler of the nineteenth-century passion for hunting, adored him: 'To see him fall! By heaven, he does even that with a grace!' True, he was inclined — according to another hunting authority, Squire Osbaldston — to puff himself up as the greatest slayer of foxes in the land, and to be savage with anyone who dared question his claim. But in the general estimation, Mr Musters exemplified all the virtues of the English squire. Even Squire Osbaldston admitted that he was 'the handsomest man of his day, of great assurance and very imposing address . . . a great favourite with the ladies'.

Rival in later life

Byron did not have a hope. He was of medium height, inclined to podginess. He limped about, dragging his accursed foot. He cut no kind of a figure on a horse. He was a lord, but could not afford the upkeep of his inheritance. And he was a boy, and Mr Musters every inch a man.

There was a tangle of history between the three families – Chaworths, Byrons, Musters. Chaworths had held lands at Wiverton since Conquest days, and in the fourteenth century acquired the manor of Annesley, between Nottingham and Mansfield. Close to Annesley was Newstead Abbey, which was not an abbey at all but a priory, and had been sold by Henry VIII to Sir John Byron for £800. Sir John was already the owner of Colwick, which his family had secured through an advantageous marriage three hundred years before. But the Byrons preferred Newstead, and around 1700 the Colwick estate was sold to a London merchant, Sir John Musters.

It was recorded of the Byron who bought Newstead that 'he loved his wife something too soon, and sought the priest's blessing on their union something too late for decorum'. The product of this prematurity was another John, known as 'little Sir John with the great beard', for the straightforward reason that he had a great beard of which he was extravagantly vain. For centuries the Byrons were among Nottinghamshire's foremost dynasties. Six of them, all brothers, fought for Charles I; the eldest of them got a peerage for his trouble, while his wife, according to Pepys, had the further honour of being the '17th mistress abroad' of the man who would be Charles II.

Newstead Abbey 1790

But something went wrong with the Byron bloodline. While the Chaworths and the Musters married sensibly, spent prudently and quietly prospered, the Byrons went to the dogs. Chief architect of this downfall was the 5th Lord, 'that devil Byron'. He inherited at fourteen, served for a time in the Navy, and when twenty-five married an heiress whom he treated abominably while frittering away her fortune. He filled the Abbey with pictures and sculptures, had a lake dug on which to stage mock naval battles and strode about the place with a pet wolf at his heels. When the money ran out, he ordered that the ancient oaks of Newstead should be felled for timber. The treasures were auctioned, followed by almost everything else – including a 'collection of animals in spirits', a handful of whips and a consignment of toothpicks manufactured in Lisbon. Shunned by his neighbours and his only son, and abandoned by Lady Byron, he retreated into the few habitable rooms left in the increasingly ruinous building, looked after by his mistress, the gamekeeper's sister, known to the locals as Lady Betty.

In his more sociable days, the 5th Lord had been in the habit of dining once a week at the Star and Garter in Pall Mall, London, with several other Nottinghamshire gentlemen. On 26 January 1765 one of the company was Byron's neighbour, William Chaworth, squire of Annesley. Fuelled by claret, an argument broke out over which estate harboured more game, Chaworth asserting that he had more on five acres than Byron on all his land.[2] After a while Chaworth left the room. Byron followed and was heard to say to him: 'I want to speak with you.' They went into an adjoining room where swords were drawn. Chaworth inflicted a minor flesh wound, whereupon Byron ran him clean through, boasting as he did so that he had 'as much courage as any man in England'. Chaworth died the next day and Byron surrendered himself for trial by his peers at Westminster Hall. He was, he said in his defence, 'an unhappy man innocent of intention, conscious only of misfortune'. Cleared of murder and convicted of manslaughter, he secured his freedom by claiming benefit of clergy and paying £30,000.

The poet Byron was dangerously fascinated by this story. When the Byron family solicitor, John Hanson, suggested that a Chaworth marriage accompanied by an injection of Chaworth cash would come in handy, Byron replied gleefully: 'What, Mr Hanson, the Capulets and the Montagues intermarry?'

His own route to the title and the ruined abbey was equally a source of delight in the way it mixed improbability, melodrama and infamy. The 5th Lord's estranged son predeceased his father by twenty years, managing to get himself killed during Nelson's siege of Calvi, in Corsica, in 1794 (where Nelson lost his eye.) Next in line would have been a nephew, Captain John Byron, known as Mad Jack and a rival to his uncle for atrocious behaviour, but he was dead, too. Mad Jack had run off with an aristocrat's wife, spent her money and seen her to her grave. Pressed for cash, he then married a Scottish girl of ample figure, Catherine Gordon, abandoned her and their son, had an incestuous relationship with his sister, and fled to Valenciennes, in northern France, where he died, possibly of consumption, possibly by his own hand.

Young Byron

Byron was that abandoned son. He was ten when he succeeded to the title, fifteen when he fell in love with Mary Chaworth. As soon as she could, she married the local sporting hero, who thereby added the estates of Annesley and Wiverton to his own at Colwick. Byron had the news from his mother: 'An expression impossible to describe passed over his pale face. With an affected air of cold-ness and nonchalance he said "Is that all?"'

For a time he sighed over her:

> Hills of Annesley, bleak and barren,
> Where my thoughtless childhood strayed,
> How the northern tempests, warring,
> Howl about thy tufted shade!
> Now no more the hours beguiling
> Former favourite haunts I see,
> Now no more, my Mary smiling,
> Makes ye seem a heaven to me.[3]

But the love was soon displaced by self-pity and curdled into some-thing close to disdain. 'She was,' he wrote later, 'my beau ideal of all that my youthful fancy could paint of beautiful . . . But I found her, like the rest of her sex, anything but angelic.'

For Mrs Musters, married life proved a severe disappointment. Mr Musters liked the ladies as much as they liked him, and when he wasn't chasing foxes amused himself by chasing other game. She became discontented and her thoughts turned back to the boy she'd brushed aside, now the most celebrated poet in Europe. She wrote to Byron, begging for a meeting: 'You will find a very old and sincere friend most anxious to see you . . . You will hardly recognise in me the happy creature you once know, I am grown so thin, pale and gloomy.'

Byron was tempted. He wrote to Lady Melbourne: 'Though it will be a melancholy interview I shall go – he [Mr Musters] has been playing the Devil, with all kinds of vulgar mistresses – and behaving ill enough in every respect.' To his wife-to-be, Annabella, he reported of Mrs Musters that 'she is separated from her husband at last after frequent dissensions arising entirely from his neglect and, I fear, injuries still more serious – at eight-and-twenty still in the prime of life – beautiful (at least she was so) with a large fortune – of an ancient family – unimpeached and unimpeachable in her own conduct – this woman's destiny is bitter'.

On further reflection, Byron decided it might be better to preserve bittersweet memories while avoiding present complications. She bombarded him with more letters, and when he was taking a break in Hastings with his half-sister, Augusta, Hanson wrote to warn him that Mrs Musters 'looks ill and talks of taking the sea air'. He fled, and never saw her again. Annabella caught a glimpse of her one afternoon outside Lady Melbourne's drawing room and reported 'such a wicked-looking cat I never saw'.

This was pure malice. Mary's own assessment of her character rings much truer: 'Soon led, easily pleased, very hasty and very relenting, with a heart I think moulded in a warm and affectionate fashion.'[4] She wrote this less than a fortnight before her death. By then, of course, Byron himself had been dead for almost eight years, his body brought back from Greece and laid to rest in the family vault beneath the church at Hucknall Torkard, a few miles from Newstead. Mary Chaworth was buried on the other side of Nottingham, at Colwick church, in the company of numerous long-gone Byrons and more recently departed Musters. A white marble figure of her was placed on the north side of the altar, but the lines written about her proved a more enduring memorial –

> . . . to his eye
> There was but one beloved face on earth,
> And that was shining on him . . .[5]

In those distant days the Trent looped north to flow past the south front of Colwick Hall. But as part of the post-war flood control programme, a new channel was dug across the bottom of this meander to hurry the river on its way. New locks were built and the old weir was replaced by a barrage of sluices to control

Colwick Hall 1790

water levels. What used to be the river is now a long, slender lake shaped like a crooked finger. The banks are studded with beds of water lilies, and swans glide across the still, dark water. Colwick Hall has been restored and functions as a swish hotel. Behind, the old Musters estate is mostly covered by Nottingham Racecourse.

Colwick church 2005

I wanted to see Mary Chaworth's memorial in the church, and measure the size of Little Sir John Byron's Great Beard. I wandered around the lake, admired the exterior of the hall, then came upon the pathetic, roofless ruin of the church, almost swallowed up by brambles, nettles and adolescent trees. Nothing but the shell survives; I learned later that the monuments had been moved some years before to the church at Annesley.

Mine was the only vessel waiting to go through Holme Lock. Immediately beside the lock, on the southern side of the river, is the National Water Sports Centre, an enormous liquid rectangle created from various exhausted gravel workings. Pencil-thin craft were knifing at enviable speed across the surface of the rowing course. Young men and women in vests and shorts were shipping and unshipping oars. Runners pounded the paved surrounds, sharing the space with cyclists bent over their machines so that, from a distance, they looked like supercharged beetles.

A little way above the lock gates there was a break in the embankment which opened into a foaming canoe slalom. A flotilla appeared from below, paddles flying. A man with a megaphone urged them on from the bank. Underneath their helmets, the faces of the paddlers were taut with effort, and none of them spared a glance for me as they thrashed past. Athleticism was all around, making me feel arthritic and past it.

I spotted the lock-keeper, a cheery fellow in shirtsleeves and crisp blue trousers. He said he could not remember having seen a punt come through before. I guided the *Otter* gently past the gates, which hissed shut behind me. We began to descend. I stood, gripping a cold, clammy metal ladder to keep the boat steady. It was like going down into an open-air dungeon. The walls were black with algae and as slippery as ice. The lock-keeper's head appeared above me, a black pebble against a shrinking sky. His voice boomed as he cautioned me to hold tight until the downstream gates were fully open. 'It can get pretty turbulent down there,' he said encouragingly. The water churned and seethed for a while, then calmed down. I rowed out into daylight and the lock-keeper waved me on my way. I asked him how far it was down to the next lock, at Stoke. 'Forty minutes,' he replied, but he must have been thinking engines.

Travelling in style

It took me two hours, battling against a mean-spirited upstream wind. The lock-keeper at Stoke had been told by his colleague to expect me. He called me Skipper and gave me a bottle of cold lager from his fridge. As I pulled away from his lock, something bad happened. Two of the three screws fixing the left rowlock to the side of the *Otter* sheared off. This had happened once before, at Shardlow, and I had replaced the screws from a meagre reserve, and repositioned the plate as best I could. It struck me now, as we drifted midstream, that unless I could get the rowlock plates securely fixed, my trip was over. The flow was now so sluggish that any upstream wind required hard rowing to counter. Without rowlocks I could not row, and the screws I had bought were clearly not up to the job. I reattached the plate with my last two screws and paddled in gingerly fashion to the settlement of Stoke Bardolph.

Apart from its pub, which was not my sort of pub at all, there was very little to Stoke Bardolph. I asked at the pub for some screws, but they were no better than mine. I tried the first cottage, where an Irish woman called Maureen said she couldn't help me but Brian next door might be able to if he was in 'because he makes boats so you'd think he'd be sure to have screws and that kind of thing'.

I banged on Brian's door and Brian appeared and I told him my pathetic tale. He went inside, then came back with a chest full of screws. He invited me to help myself. I explained, shamefaced, that my screwdriver was almost as inadequate as my screws. He expressed mild surprise that I should have embarked on such a venture without having paid more attention to my fixtures and fittings.

In the end we agreed that – with him being a professional builder of narrowboats and me being something else – it might be better if he did it rather than me. He inspected my rowlock arrangement and said: 'Your screws are completely useless.' Well, by then I could have told you that. He produced a handful of tungsten-tipped, silicon-lubricated screws. 'These should do the job,' he said. After some whirring of his battery-powered screwdriver, my rowlock plates were snug against the *Otter*'s sides, immovably fixed. I felt like weeping with gratitude.

Brian then looked at the rowlocks themselves. Why, he wondered, did they not have a pin through the hole at the bottom to stop them jumping out and disappearing overboard. I said I had meant to do something about that but hadn't quite got round to it. 'Got any spares?' he asked. One, I said. 'Good thing you didn't lose two, then,' he commented. Then he clipped a metal stopper into each of them. 'Should be all right now.'

I tried to persuade him and his wife to be my guests at the pub, but he declined. He seemed to want no other reward than to have saved an incompetent from certain disaster. Later I rowed across to a field and put up my tent on a grassy ledge occupied by a community of Canada geese. Ignoring their honking and outbursts of pointless wing-beating, I sat drinking the last of my red wine, speculating on what the odds might be against anyone having a stroke of luck such I had just had. I was now two-thirds of the way there, and I felt a surge of confidence that I would make it.

Chapter 15

Poor Lame Boy

I was woken up by the snapping of tent flaps in the wind. I looked out at a slate grey sky. The river beneath it was wide and black, its surface chopped and grated into nasty little waves, advancing upstream with fixed purpose. If my sleeping bag and airbed had been more seductive I would have retreated and shut the world out. But they weren't, and I had a way to go.

The wind was blowing from the north-east, not violently but with an unflagging persistence, like a wife or husband who just will not let you be. It was a swine to row into. The *Otter*'s lightness, in most respects her finest virtue, was here her flaw. The wind was forever thrusting underneath her square, angled prow, lifting it and pulling and pushing it one way and the other; while the waves slapped at it, slap, slap, slap, doing their utmost to drive me and her back to Nottingham.

Flat, featureless farmland stretched away on both sides, offering no protection from my enemy. Out in the middle the motion alarmed me, so I tried to hug my bank. But another danger lurked. It was Saturday, the working man's day of leisure, and the anglers were out in force. Some were tucked away in little tents, with their rods outside, laid horizontally on rests with electric alarms attached. Others were wrapped up on recliners, muffled against the wind, thermoses at their sides. Every so often one would get up, revealing rubber thigh waders beneath jacket or fleece. He would pick up a rod, reel in, inspect and replace the bait, and wade out a couple of yards. The rod would bend like a bow under forward propulsion and moments later there would be a splash far out as the weight hit the water. It occurred to me that being struck on the head by a one-and-a-half-ounce lead flying as fast as a bird might kill me.

I heaved on the oars for all I was worth, blessing my saviour at Stoke Bardolph with each stroke. If I let up for a moment, the *Otter* immediately stopped; and if I rested for ten seconds, we started scooting backwards. At Burton Joyce, which is about a mile below Stoke Bardolph, the river swung from the north to the east. The wind went with it. Normally I would have been moderately interested to know that the Joyce in Burton Joyce was derived from Jorz, the bigwigs here in the thirteenth and fourteenth centuries, one of whom is alleged to have been a pupil of the great mage, Albertus Magnus, and two of whom were certainly Archbishops of Armagh. I might also have stopped to mooch around Shelford Manor, which is on the opposite bank and was the scene of a vicious little engagement in the Civil War, ending with its Royalist commander, Colonel Philip Stanhope, being stripped naked, thrown on a dunghill and killed.[1] But on this morning I was interested only in beating the wind.

At around nine o'clock I struggled into Gunthorpe Lock. I tied up at the bottom of the lock cut and turned on my radio to listen to the first rugby union test between the New Zealand All Blacks and the British Lions. Almost at once the Lions captain, Brian O'Driscoll, was carried off on a stretcher. The Lions were crushed. My mood, already sombre enough, deepened into heavy gloom. Sitting on a plank in a boat on a chilly, dreary morning, eating a breakfast of stale bread and jam, listening to a tale of calamity from the other side of the world, I felt quite sorry for myself.

Below Gunthorpe the river resumes a north-easterly course, squeezed against the foot of a long ridge known as Trent Hills, which extends five or six miles along the east bank. On the west side the land is flat, and a good deal of it – gravel laid down in the river's prehistoric youth – has been devoured by diggers and removed by trucks to hold together motorways, bridges, factories and assorted public buildings.

Out of sight on top of the ridge is the Fosse Way, which was built by the Romans to enable legionaries and tax collectors to get around between Bath and Lincoln. For most of its two-thousand-year history it connected one quiet part of England with another; now it is the A49, a roaring, snarling strip of mechanised din. A little way back from it on either side are the villages of Car Colston – home of the admirable Thoroton – and East Bridgford and Scriveton, where the regicides Francis Hacker and Edward Whalley came from. Hacker took his medicine like a man when the time came, offering his neck to the noose with the words: 'I have no more to say for myself than that I was a soldier under command and what I did was by the commission you have read.' Whalley fled to America when he saw that the old, monarchical ways were staging a comeback, disappeared into the woods of New England and was never heard of again.

Ferry at Hazelford early 1920s

A cheery soul travelled this stretch of river in the cheerier times before Charles Stuart turned the country sour. The sight of Flintham Wood, which overlooks the river at Hazelford, spurred him to compose one of the better specimens of Trent verse:

> The ground we trod was meddow, fertile land,
> New trimmed and levelled by the mower's hand.
> Above it grew a roke, rude, steepe and high,
> Which claims a kind of reverence from the eye:
> Betwixt them both there glides a lovely stream,
> Not loud but swift; Maeander was a theme,
> Crooked and rough, but had the poetts seen
> Straight, even Trent, it had immortall bin.
> This side the open plain admits the sun
> To half the river; there did silver run;
> The other half ran clouds; where the curl'd wood
> With his exalted head threatened the flood.

The poet was a clergyman, Richard Corbet, at various times Dean of Christ Church, Oxford, Bishop of Oxford and Bishop of Norwich, and a favourite of both James I and Charles I. He was liked not so much for his piety, which was average, but for his jollity, which was exceptional. John Aubrey has a story in *Brief Lives* about him drinking one day with his friends in a tavern in Abingdon. It was market day and they were watching a singer of ballads who complained that his trade was poor, whereupon 'the jolly Dr putts-off his Gowne and putts-on the Ballad-singer's Leathern jacket and being a handsome man and had a rare full voice he presently vended a great many and had a great audience'. According to Aubrey – who was not always particular about sourcing his stories – Bishop Corbet's wife was 'a very beautiful woman whom 'twas said he bigott'.

I regret to say that the charms of Flintham Wood were pretty much lost on me. Eight hours of hard rowing had taken their toll of shoulder joints, arm sinews, back muscles and buttocks, leaving me feeling somewhat jaded and very much in need of rest and nourishment. The one brief distraction from my discomfort came when we passed from sheet 129 of the Ordnance Survey map — my companion and guide since I shot into King's Mills — to 120, which showed the wriggling way to Newark. The map gave me hope. Around one more bend was a place called Fiskerton, and very close to the river were the letters PH. Public House.

It would be hard to exaggerate the importance of pubs to me on my journey. Forget hard. Impossible. No pilgrim hauling himself through the streets of Santiago de Compostela towards the cathedral, or lifting his eyes to behold for the first time the al-Haram mosque in Mecca could have experienced a more exhilarating lifting of the spirits than I did when I came within sight of the Bromley Arms in Fiskerton. It was well placed to inspire the water traveller, a whitewashed building standing welcomingly beside the wharf which reared above the river. I tied up to the wooden landing stage, tottered up the stairs and into the bar. A minute later I was downing a first pint of Hardys and Hansons bitter. PH, I thought. What a fine union of letters!

Fiskerton had a shop as well. When I was stronger I reached it and bought a newspaper, chocolate, eggs, milk, sausages and two bottles of serviceable Rioja. I had already decided to go no further that day. Fiskerton was the place for me.

The Bromley Arms didn't do rooms, so I paddled across to the far bank, where a long, straight farm track came down to the river. I put up the tent close to a bay willow, among clumps of comfrey and white clover. Behind me stretched fields of beans, wheat and oilseed rape; in front of me the river slid by. It was a good spot to be. Later, I recrossed the water to the pub where I ate a plate of liver, bacon, mashed potatoes and thick, brown gravy, and drank four, possibly five, pints of bitter. By bedtime I felt like Bishop Corbet.

* * *

One hundred and fourteen years after Byron's body was interred beneath Hucknall Torkard church, a ceremony took place there whose grotesque and macabre nature must have appealed to the poet if he'd been there to witness it. The vicar, Canon Thomas Barber, had got it into his head that it was his duty to investigate persistent rumours that Byron's body had been spirited away. He applied to the Home Office and to the 10th Lord Byron – who happened to be Vicar of Thrumpton – for permission to enter the vault and open the coffin. Canon Barber protested that he was moved by a desire 'to establish some archaeological points of general interest', and not by 'morbid curiosity', so one assumes that being morbidly curious must have had something to do with it.

A detailed description of what happened is included in Fiona MacCarthy's biography of Byron.[2] Witnesses testified that his body was in almost perfect condition. His head, torso and most of his limbs were solid, and his hair still curled around his brow, although it had gone quite grey. The church warden, Mr Houldsworth, noted that 'his sexual organ showed quite abnormal development'. There was one odd thing, which was that his right foot was detached and lay at the end of the coffin.

That foot! No appendage in literary history has received more attention. Byron himself blamed his mother for his deformity, claiming it was due to her having worn excessively tight corsets during her pregnancy; but then he blamed most things on her, some with justice, a great deal not. In fact, it was almost certainly a genetic defect, but, whatever the cause, it gave him reason to believe that he was not as others, and not whole.

He was touchy about it from the start. The story goes that when he was being taken out by his nurse in Aberdeen, another nurse commented: 'What a pretty boy Byron is! What a pity he has such a leg', whereupon he struck at her with his little whip shouting, 'Dinna speak of it.' In Nottingham he was taken to a sawbones called Lavender who tried to correct the fault by twisting the foot around and screwing it up in a wooden box of his own design. This, and other treatments, did no good, and simply caused pain and humiliation. According to Edward Trelawney, who was with Byron in Greece towards the end, and wrote several highly coloured accounts of their friendship, 'his deformity was always uppermost in his thoughts and influenced every act of his life, spurred him on to poetry as that was one of the few paths to fame open to him'.[3] Poetry was one way. But there was another element in which he could excel, where his other physical attributes more than compensated for what was incomplete. 'He was built for floating,' Trelawney recalled, 'with a flexible body, open chest, broad beam and round limbs.' Byron had a passion for swimming that easily spilled into obsession. He swam in the lakes at Newstead, he swam the Thames through London and the Cam through Cambridge. Wherever he was, he looked for water and someone to challenge, alive or dead.

On his first escape from England, he conquered the winds and the tides to cross the mouth of the Tagus in Lisbon. Cruising Asia Minor in the frigate *Salsette*, Byron conceived the urge to match the legendary Leander by swimming the channel between Sestos and Abydos known as the Hellespont. The first time he and Captain Ekenhead, a lieutenant on the *Salsette*, tried it, they were beaten by the cold and the strength of the currents. Byron went into intensive training, swimming daily around the castle guarding the Dardenelles. He and Ekenhead tried again, successfully this time, and Byron wrote to a friend: 'The immediate distance is not above a mile but the current renders it hazardous, so much so that I doubt whether Leander's conjugal powers must not have been exhausted.' Leander, of course, had swum it both ways every night, pleasuring Aphrodite's priestess, Hero, in between. But then he had drowned, and was a legend anyway; whereas Byron was real and had survived. 'I plume myself on this achievement,' he crowed, 'more than I could possibly do on any kind of glory, political, poetical or rhetorical.'

No one could charge Byron with being unduly modest, or with

being either a gracious winner or a good loser. Trelawney once beat him in a swimming race and Byron sulked like a spoiled child. When living in Venice — a period chiefly devoted to the exercise of his abnormally developed sexual organ — he made the acquaintance of a soldier, Angelo Mengaldo, who had fought for Napoleon and distinguished himself by swimming the Danube under enemy fire. One evening he and Byron were boasting of their exploits, and Byron challenged the cavalier to a contest.

It took place in June, when the water had warmed up. There were three contestants: Byron, Mengaldo and a young Englishman called Alexander Scott. The course was from the Lido past San Giorgio Maggiore into the Grand Canal. The Italian was soon floundering. 'Scott and I beat him hollow,' Byron reported with glee, 'leaving him breathless, miles behind and knocked up, hallooing for the boat.' Scott got as far as the Rialto, while Byron kept going to the point where the Grand Canal opens into the lagoon, altogether four and a half hours in the water. He bragged at length about his prowess that day in the water and the bedchamber: 'I could not be much fatigued having had a *piece* in the forenoon and taking another in the evening . . .'[4]

Young Byron

According to J. B. Firth, author of the *Highways and Byways of Nottinghamshire*, Byron used to swim and go fishing in the River Greet, which enters the Trent at Fiskerton. He was then living and quarrelling with his mother in Southwell, which is three or four miles across the fields from Fiskerton. Mrs Byron had rented Burgage Manor, a spacious stuccoed house looking out on to Burgage Green, on the edge of town. Byron, typically, did not have a pleasant word to say about the place – 'this crater of dullness . . . cursed, detestable and abhorred abode of scandal . . . all cards and old maids', he called it.

In fact he amused himself well enough when back from school or university. He took part in theatricals at the Assembly Rooms, and performed duets with his sensible friend, Elizabeth Pigot, who lived across the green and remembered him as 'a fat, bashful boy with his hair brushed straight back'. He conducted flirtations and addressed verses of a swooning character to various girls, then collected them and some others into a volume which he called *Fugitive Pieces* (his friend, the Reverend John Becher, objected to a reference to the poet panting in his mistress's arms as being 'rather too warmly drawn', whereupon Byron had all the remaining copies destroyed).

Another set of verses – *The Adieu* – addressed the Greet:

> Streamlet! Along whose rippling surge
> My youthful limbs were wont to urge
> At noontide heart their pliant course
> Plunging with ardour from the shore . . .

The poem was given the subtitle *Written Under The Impression That The Author Would Soon Die*, and one wonders how high his reputation would have stood had his premonition come true. As for the streamlet, looking at it today it's difficult to imagine anyone swimming in it, let alone being moved to write a poem to it. Dredging work carried out to prevent flooding has wrecked it, reducing it to a dark dribble that creeps along between high, rounded banks as if it were carrying some shameful secret. Incredible to think that this was once a trout stream, on which a friend of the 'Trent Otter' took fifty-eight trout in a day, 'many of the fish running up to two-and-a-half pounds'.

I hate the sight of a ruined river. I followed it up to Fiskerton Mill, which was still working when Peter Lord wrote his book about the Trent forty years ago and where there is now hardly enough flow for a game of Pooh Sticks. Higher up, the stream crawled around the edge of Southwell Racecourse, a murky, furtive, broken thing. I couldn't bear to look at it any more, and it was a relief when the path took me away from it.

Southwell Minster

But it would have been difficult to stay dejected for long on such a morning. The weather had changed again during the night. Some subtle realignment of cold and warm fronts had pulled the air stream away from the chilly North Sea into the south. When I squirmed out of my tent, the dew was sparkling, and the surface of the river was like glass under a blue sky brushed with cirrus. By the time I reached the outskirts of Southwell it was a brilliant summer's day, and the sight of its Minster's great square towers and shining roofs put the spring back in my step.

The Minster again

I came into the town along a path shaded by what I took to be beech trees, thick with clusters of yellow flowers. Each tree was emitting a loud droning, and when I peered at one closely I realised the sound came from thousands of bees which were sucking themselves stupid on the nectar from the flowers.

For a time I sat outside Byron's house, picturing the fat boy poet trotting across to take tea with Miss Pigot before rehearsing their rendition of 'When Time Who Steals Our Years Away'. Then, as the Minster bells rang out, I passed along streets of pretty red-brick cottages and elegant town houses to the great church. Inside, a train of choristers was waiting to process into the chancel. Behind the boys were lined the deacons and archdeacons, resplendent in green; and behind them a bishop, in his mitre and raiment of gold, grasping his staff. The organ thundered and the voices came together in the Kyrie Eleison, the sound welling up to the sunlit emptiness beneath the roof. I felt stirred, uplifted, a prickling in the pit of my stomach, a moistening around the eyes, as if I was being reclaimed by my church-going past.

I would have joined the service had it not been for my shoes.

I had bought them from a specialist outdoors shop, and although they were extremely comfortable and robust, there was something in the material that made them smell like pig slurry when they were damp. I did not want to cause offence to the regular worshippers, so I stood outside the back of the chancel. Inside, the bishop embarked upon his sermon. It was three weeks before the leaders of the G8 countries were due to gather in Scotland to discuss how they were going to grapple with the great issues of world poverty and global warming. In the event, the bomb attacks in London on 7 July rendered the summit even more meaningless than it would otherwise have been, but at this stage the bishop clearly felt that it was his duty to help concentrate the minds of the leaders.

As the platitudinous waffle flowed mellifluously forth, I went out to look at the astounding carvings that adorn the exterior of the Minster and its adjoining Chapter House. Then I set off to investigate a question that had been nagging at me: could Byron have eaten one of the very first Bramley apple pies?

The genesis of a new variety of apple is usually the result of laborious and virtuous experimentation, as with the Cox's Orange Pippin reared by the laborious and virtuous retired brewer Richard Cox near Slough in the 1820s. But just occasionally extreme serendipity comes into play. A pip from an apple is like an egg from one of us. It carries genes derived from its parent, but they are arranged in a manner unique to itself. To secure a Cox's Orange Pippin tree, you must use a graft from another Cox's Orange Pippin tree. How a pip fallen by accident on fertile ground will turn out is anyone's guess.

Some time in the first decade of the nineteenth century, in Southwell, a girl called Mary Ann Brailsford picked up a pip from an apple her mother had cut up and planted it in a flowerpot.[5] It germinated, grew too big for its pot and was put out in a corner of their garden in Church Street. In time the little tree began to produce large, green apples of an irregular shape. They were no good as eaters, but when cooked, their snow-white flesh reduced to a pale, fluffy purée with a powerfully acidic flavour that could hold its own against both sugar and spice.

The Brailsfords lived in Church Street for many years, but the house was eventually sold to a local butcher and publican, Matthew Bramley. One autumn day in 1856, a local lad, Henry Merryweather – who worked with his father at a nursery nearby – met another gardener who was carrying a basket of splendid apples. He told young Merryweather that they came from a tree grafted from Mr Bramley's tree. Being as sharp as the flavour, Merryweather spotted an opening and, within seven years, Merryweather's Nursery was offering the Bramley's Seedling for sale.

The quite marvellous thing is that the original tree, sprung from Mary Ann Brailsford's pip, is still standing. Actually, standing isn't quite the word, as it was blown down in a storm a hundred years ago. But it is still growing from a recumbent position, and it is still producing apples. I know this because I was shown it by a man with a beard called Coulson Howard – 'please make sure you get that the right way round,' he said in a very particular manner, so I hope I have. Mr Howard is a nephew of the current owner of the tree and was happy to usher me into its presence.

Original tree

To anyone interested in apples, as I am, the occasion was invested with almost sacred significance. This, after all, is the undisputed champion of cooking apples, the source of all the Bramley's Seedlings across the world. Despite its great age, it was looking pretty good. A great cloud of foliage filled the corner of the garden, studded with clusters of the new season's fruit, still small and hard but promising another abundant crop come October. Peering into the shadows, I made out a trunk rising at an angle from the gnarled, hollow, desiccated remains of the fallen ancestor. What was amazing was that such vigorous growth should have sprouted from what looked like a corpse of bark and brittle wood.

I thanked Mr Howard and congratulated him and his family on having looked after the treasure so tenderly. As I trudged back towards Fiskerton, I considered the chronology of the Byron/Bramley question. Byron and his mother moved to South-well in 1804, and she kept the house until 1807. Thereafter, until 1814, Byron's home was Newstead, which is not so far from South-well. He remained on friendly terms with Miss Pigot and her family, and it is entirely possible that, when revisiting old haunts, he might have been offered a slice of apple pie ('good apple pies', wrote Jane Austen, 'are a considerable part of our domestic happiness').

But what of Mary Ann Brailsford's pip? According to Peter Lord's book, this was planted in 1805, which would allow plenty of time for the poet to have encountered the fruit. Unfortunately most other authorities put the date at 1809, which – allowing, say, four years for maturing and fruiting – is probably cutting it too fine. The strong probability, therefore, is that one of the few pleasures of the senses that Byron never experienced was to taste the nonpareil of cooking apples; which is a shame, for had he done so, might not that exceptional symbiosis of sweetness and sharpness have roused the Muse?

I fumbled with some lines of my own:

> In Betty's garden
> Grew a tree
> Bright with apples
> A sight to see.

The rhymes began to give me trouble – 'wheedling' for seedling, 'family' for Bramley seemed to promise little. I gave up and walked on under the hot sun.

Chapter 16

The Impostor

Track at East Stoke

The track ran in an almost straight line from the river, past my tent, to the foot of the ridge carrying the Fosse Way. There it turned abruptly left, next to a heap of mangel-wurzels – or possibly swedes – that had been harvested from the adjoining field and left to await their fate. Further on it turned right, past the Church of St Oswald and the gates of the big house, Stoke Park, towards the village of East Stoke.

The church itself is not very interesting, but I was much taken by a memorial outside it to Lord Pauncefote, who was Britain's first ambassador in Washington, where he died in 1902 from the effects of a prolonged attack of gout. One of the Beatitudes – 'Blessed Are The Peacemakers' – is inscribed on the tomb, above which rises a bronze statue, now seasoned to a bilious green, of winged and bare-breasted Athene, who is holding an olive branch.

Peace

Peacemakers and olive branches were conspicuously absent from the ridge above Stoke Park and the meadows below it on 16 June 1487. Blood, guts and the clash of weaponry were the order of that savage day, and it's doubtful whether even Lord Pauncefote's diplomatic skills could have restrained the White Rose of York from having its final, futile tilt at the Lancastrian usurper, Henry Tudor. The Battle of Stoke Field was fated to happen. With it, a bitter chapter in history closed.[1]

To those gathered there that summer's morning, it must have seemed a complete mismatch. Henry, who had marched downstream from Radcliffe, had forty thousand men, perhaps more. Ranged against them along the slope was a rebel force of about eight thousand, comprising several companies of Swiss and German mercenaries, a mob of wild Irish volunteers and a thousand or so English recruits picked up on the long march from the landing on the Lancashire coast. In command, theoretically, were the Earl of Lincoln, nephew of the deposed Edward IV and nominated heir of the defeated and dead Richard III; and Francis, Lord Lovell, who had fought beside Richard at Bosworth Field. But both were military novices and content to defer to the leader of the continental mercenaries, Martin Schwartz, who had fought more battles than he might care to remember.

Henry delivered the customary address to the troops, blaming everything on Lincoln and Edward IV's meddling sister, Margaret of Burgundy, before retiring to a vantage point, possibly the church tower at Elston, to watch the proceedings. As his vanguard, led by the Earl of Oxford, advanced, his archers let fly a cloud of steel-tipped arrows into the rebel ranks. Casualties were high among the Irish tribesmen, who wore no armour and depended on daggers and javelins. Schwartz and Lincoln decided that they must attack or perish where they stood, and they urged their men down the slope. The mercenaries, with their pikes, halberds and arquebuses, were justly feared for their fighting prowess and discipline; while at close quarters the ferocious, half-naked Irish were an alarming proposition.

For a desperate few minutes the rebels seemed to have the advantage. Henry's thoughts may have gone back to Bosworth, where Richard's overwhelming numerical advantage had counted for nothing once the tide turned against him. But now Oxford rallied his men. The vanguard, backed by archers and cavalry, pressed the rebels back. Schwartz, Lincoln, Lovell and the other commanders strove to stage a rally, but the pressure was too much and the line broke.

There was carnage on the ridge, and carnage on the sloping ground and flat fields leading to the Trent. Schwartz, Lincoln and the Irish commander, Lord Geraldine, were cut down and their bodies displayed. In all, it's likely that half the rebels – four thousand men – were killed. Henry left the corpses to be disposed of and marched to Newark, where he gave thanks in the Church of St Mary Magdalene and dedicated his banner to her. The next day he marched to Lincoln, where a selection of rebels were hanged and the surviving mercenaries were told to go home. That night a feast of celebration was held, with geese, sheep, oxen and pike on the menu. As he tucked in, Henry had much to be grateful for; much to think about, as well.

The Battle of Stoke Field finished off the Yorkist military threat for good. But it left unresolved issues and two lingering mysteries. The minor mystery concerned the fate of Lord Lovell, who was variously reported to have been killed on the battlefield, to have fled to the river and been drowned trying to cross it, and to have got clean away. Francis Bacon referred to rumours that he managed to reach his family seat at Minster Lovell in Oxfordshire, where he hid in a secret vault.[2] Twenty years later, so the story went, his skeleton was found sitting at a table with books, paper and pens in front of it; the presumption being that whoever was supposed to look after him had died or run away or been arrested or gone mad or simply got tired of it, leaving the poor fellow to starve to death, his cries unheard.

Henry Tudor would not have lost much sleep over Lovell. But the other mystery was a different matter, and a mystery it remains to this day – although if anyone knew the answer to it, it must have been Henry himself. But if he did, he kept it to himself. Even now, no historian can with certainty answer the question: who was Lambert Simnel?

The conspiracy that was crushed beside the Trent was an amazingly tortuous affair. The official version, sanctioned by Henry, was written by an Italian cleric called Polydore Vergil, who never went near the place, did not even come to England for the first time until fourteen years after the battle, and appears not to have interviewed any of the surviving witnesses. Nevertheless his account, with a few adjustments, acquired in time and through repetition the status of fact.

Vergil stated that the plot was dreamed up by a humble priest living in Oxford, Richard Simon (or Simons), who yearned for greatness and thought he'd found a short cut to it. Henry had taken the throne by force and, judged by the criteria of legitimacy, his claim was inferior to that of at least three others. Two of these were the Princes in the Tower, the sons of Edward IV; they had allegedly been murdered in 1483 on Richard III's orders, but their bodies had never been shown and rumours abounded that they'd been spirited away. The third was the Earl of Warwick, the young son of Edward IV's brother, the Duke of Clarence. If any one of these could be produced in public, and the force summoned to challenge Henry's, then another revolution was on the cards. There was no shortage of ambitious, bitter, vengeful, greedy men ready to play those cards.

By his own doing, claimed Vergil, Richard Simon chose a boy who would be king. He was called Lambert Simnel, the son of one Thomas Simnel, and the priest coached him in Latin, French, his assumed past, and courtly manners. Apparently because of the persistent whispers that the Earl of Warwick had been set free from the Tower, Simon decided that this was the best guise. It was as the Earl of Warwick that the boy Lambert was displayed in Dublin early in 1487, and there hailed as Edward VI and crowned as such on 24 May.

Henry tried to nip the plot in the bud. He had the 'real' Earl paraded through the streets of London, and had circulated the details of the conspiracy, obtained – so it was said – from the priest Simon, who had been arrested and had, so it was said, confessed all. But it was too late. Lincoln and Lovell had been at work on the Continent for months recruiting troops and planning an invasion. A military showdown was inevitable.

In Vergil's account, both priest and impostor were at the battle, and both survived, and a wise and generous King Henry decided to spare their lives. The priest was imprisoned and never heard of again. The impostor was taken into the royal household and employed as a spit-turner in the kitchens and subsequently as a falconer. Vergil says he was still alive in 1513, and there is a record of a 'Lambert Symnell, yeoman' attending a funeral in 1525.

This pretty much became the authorised version, with an additional twist – provided in a life of Henry written by a French friar, Bernard André – that Simon had begun by passing the boy Lambert off as the younger of the Princes in the Tower, and had decided late in the day to switch mounts. It is as watertight as a sieve.

For a start, there is the name. Close scrutiny of contemporary rolls shows that – apart from the one Lambert Simnel – the Christian name was almost unheard of, and the surname entirely so. Furthermore, a brief record of the Battle of Stoke Field compiled at the time or not long afterwards by an anonymous royal herald refers to the capture of the Pretender, 'whose name was indeed John'. The name Lambert Simnel was not made public for the first time until the conspiracy was reported to Parliament six months *after* the battle. Professor Michael Bennett concludes in his book *Lambert Simnel and the Battle of Stoke* that 'the suspicion must be that the whole name was an invention'.

Then what of the priest? Vergil identifies him as Richard Simon, yet the records of the Convocation of Canterbury for February 1487 – when he allegedly confessed – refer to him as William. And how come he was spilling the beans in February yet free to be present at the battle in June? Professor Bennett's suggestion that there may have been two brothers, one called William and one called Richard, both priests, both conspirators, both of whom disappeared off the face of the earth after the battle, seems a touch fanciful.

Francis Bacon, who endorsed the 'double imposture' version,

offered a plausible explanation of Henry's decision to spare the impostor's life and to keep him close by. 'If he suffered death,' Bacon wrote, 'he would be forgotten too soon, but being alive, he would be a continual spectacle and a kind of remedy against the like enchantments of people in time to come.' Henry seems to have used him as a court joke as well. When he entertained a group of Irish lords to dinner, he informed them that 'their new King Lambarte Symnelle brought them wine to drink', and cautioned them to be careful 'lest you will crown apes at length'.

But neither Bacon nor anyone else could paper over the other obvious discrepancies and absurdities. Why was the captured priest not examined and a full record of his treachery made public? And why was he then so thoroughly erased that no one would ever be able to question him again? Why didn't Polydore Vergil and Bernard André take advantage of Lambert Simnel's presence at court to ask him about his origins and actions? Or, if they did, why was so little information about his past revealed? Why did Henry, having given strict instructions that the Earl of Lincoln should be taken alive at the Battle of Stoke Field, not punish those who killed him? How was it that all the accounts suggest an age of sixteen or seventeen for the impostor, yet the Act of Attainder passed by Parliament in November refers to him as 'Lambert Simnel, a child of ten, son of Thomas Simnel, late of Oxford, a joiner'?

The whole thing reeks of cover-up. The obvious assumption must be that, since Henry Tudor was the chief beneficiary, the cover-up was ordered by him or on his behalf. But why was a cover-up deemed necessary? And what was covered up?

Here is one persuasive explanation.[3] Suppose that the youth paraded as Lambert Simnel after the Battle of Stoke Field was not the same as the one acclaimed as Edward VI before it – in other words, that a substitution was made on the field of battle. Suppose that the pseudo-Edward VI was actually killed in the fighting. It would have suited Henry to be able to produce him afterwards to demonstrate that he was a nobody; and to keep a very close eye on him thereafter to ensure that no one got a chance to examine him or his story too intently. The same would apply if the pseudo-Edward VI was not killed but captured, *and his true identity then established*. The obvious move then would have been to dispose of him secretly and produce the substitute.

I like the second of these hypotheses. It only convinces if the captured pseudo-King did indeed have royal credentials that would have represented a continuing threat to the new Tudor dynasty. He could have been the Earl of Warwick, in which case the Earl of Warwick shown off in London that February was another fake. He could have been the younger of the Princes in the Tower, Richard Duke of York, in which case the murderer was Henry and not hunchback Richard III. It's even been suggested that he was the elder of the princes, in which case Henry was guilty of double villainy.

If a substitution was made, who did the deed? The herald's report offers a possible clue: 'And there was taken the lad the rebels call King Edward, whose name was indeed John, by a valiant and gentle squire of the King's house called Robert Bellingham.'

Valiant and gentle Robert Bellingham may have been, but he was in deep trouble not long after.[4] He'd had his eyes on a rich widow, Marjery Ruyton, who was living with her father near Solihull. With the assistance of a band of companions, Bellingham burst into the house, assaulted the father, threatened the other members of the household that 'he that stirreth shall die', and forcibly carried the widow off. Nine days later he and his accomplices were arraigned before the justices in Warwick. Yet a few months after that, when the case came up at the Marshalsea in London, the prosecution offered no evidence. Bellingham was not merely set free with permission to marry Widow Marjery, but given a job as bailiff on one of Henry's estates.

Now who doesn't smell a rat?

* * *

The landscape of the battlefield and the surrounding countryside has changed beyond recognition since that June day half a millennium ago. Then the fields were cultivated according to the medieval strip system, and were open and hedgeless. The village of East Stoke was much bigger, extending well down the slope towards the river. The Trent itself was split into two or three much shallower channels, with much lower banks and much more meandering.

Sitting in the shade of the willow, eyes on the water, I tried to picture the scene as the rebel force came clattering down the lane from Southwell to Fiskerton ford. I pictured the Yorkist standards fluttering, the white rose proud, the summer sun glinting on the weaponry – 'swerdys, speres, marespikes, bowes, gonnes, harneys, brigardynes, hawberks, axes and many others', as the herald listed them. I saw the air thick with dust from the hooves of the cavalry. I heard the snorts of the horses mixed with the Gaelic cries of the Irish tribesmen, the gutteral shouts of the Swiss and German mercenaries, the cultivated tones of Lincoln, Lovell and the other English nobles. I watched them into the stream, the water at midsummer flow, up to the horses' knees and the infantrymen's waists, and across and over the open ground towards the ridge, hope and dread churning in their breasts.

And somewhere in their midst, a boy wearing a crown, 'a comely youth', as Polydore Vergil wrote, 'and well-favoured, not without some extraordinary dignity and grace of aspect'; the son, perhaps, of an organ-maker, a joiner, a baker, a cobbler; or perhaps of a duke, or even a king. Whoever he was.

Chapter 17

Inns, Castles and Salmon Tails

At lunchtime the Bromley Arms was invaded by a force of red-faced Fun Runners, their vests damp and dark with sweat. I was glad I had nothing to do with it. The heat was fierce and the sun high as I eased away from the pontoon and rowed past the mouth of the wretched little Greet. To get back to a north-easterly direction, the Trent here executes a south-easterly loop in the shape of a fish-hook, with the pleasingly named Gawburn Nip as its shank and Gawburn Holt as its gape. It then winds through flat countryside to Farndon, where there are sailing clubs, a marina and a riverside pub. The place was seething with boaters, drinkers, sunbathers, scullers, narrowboat people, dog-walkers, stick-throwers, strollers, river-watchers, pleasure anglers. I glided past, my straw hat pulled down against the glare, my rowlocks creaking reassuringly.

Further on, the view was overwhelmed by the wasteland surrounding what had once been Staythorpe Power Station. Its cooling towers had gone, and without them there was nothing to draw the eye away from the desolation of abandonment: slabs of concrete, drifts of rubble, rusty cabling left twisted at the foot of rusty pylons, doorless buildings, clumps of useless transformers like misshapen, ferruginous fungi, tarmac split by eruptions of thistles and brambles, the wilderness patrolled by bounding hares.

There was one visual distraction, inside the high fence separating this tract of dereliction from the river. A gigantic concrete

Power in trust

bunch of what looked like severed electricity cables reared at an angle from a crumbling square plinth. Even I could tell it was a work of art: a sculpture, to be precise, commissioned in a bygone age by the old Central Electricity Generating Board to stand between the original Staythorpe A installation, and the newer, bigger, more powerful Staythorpe B. It was executed by Norman Sillman, then head of sculpture at Nottingham Art College and perhaps better known as the designer of many recent coin faces. Entitled *Power In Trust* — which was the motto of the CEGB — it was apparently intended to represent a turbine resting between boiler tubes and to symbolise the vision of the new power-driven, technology-based industrial golden age which some believed was dawning.

That was less than fifty years ago. Now Staythorpe A has gone and Staythorpe B, and there is no Staythorpe C. The CEGB itself seems to belong to an age as far removed from ours as that of Byron. Staythorpe's future is in the hands of a company which calls itself 'RWE npower', and says it wants to build a gas-fired generating plant there. In the meantime, all that's left is a lump of old concrete mouldering away next to the perimeter fence, and an ungodly mess all around.

The Trent divides here. One arm, classified as unnavigable, plunges over a toothy weir and bypasses Newark to the west. The other proceeds in an orderly manner into Newark, through the centre of town, and out the other side.

The story of this parting of the ways is old and somewhat complicated. What is now the western arm was created sometime in the thirteenth century when the local nobs, the Suttons, had a cutting dug from the main river into a brook running through their land to power their mills. In time this became the principal channel, which caused the citizens of Newark to complain that they had been cheated of their river. Eventually, in the eighteenth century, the town burgers obtained permission to tidy up the junction between the secondary Trent channel and the River Devon (which is pronounced Deevon and wriggles up from Leicestershire) to create a reliable waterway through the town, with weirs, locks and a towpath for the horses to haul the barges.

These various interferences caused the Trent as it had been for millions of years to give up the ghost.[1] 'Antiently', one of the histories of Newark recorded, 'the River Trent passed near the town about 345 yards distant from the Castle . . . The bed of the old river is very apparent and to this day is called the Old Trent.' And to this day, for the name is on the OS map, next to a fine, wandering line of blue going nowhere.

I had already decided that my way should be the way over the weir and down the supposedly unnavigable branch, because I'd had a look at it and it was manageable and it would be my last chance of a little adventure. But I also wanted to see something of Newark. And even more than I wanted to see Newark, I wanted to sleep in a bed rather than a bag, and stand beneath a shower to wash away the sticky, sweaty, smelly dirt that seemed to have formed a second skin around me since my stay in Swarkestone four nights before.

The lock-keeper at Newark Lock advised me against leaving the *Otter* on view along the Town Wharf. 'Yobboes,' he said angrily. 'If they can't steal it they smash it up.' At his suggestion I made for the British Waterways marina, a big pond off to the left, entered beneath an elegant blue footbridge. It was crammed with cruisers, narrowboats, dinghies, rowing boats – every kind of river craft bar punts. Feeling rather conspicuous, I paddled around a line of pontoons searching for a space. It was well after six o'clock, but the sun was still beating down. Men in sunhats and shorts were crouched in their boats, painting, polishing, varnishing, sand-papering, in some cases just feeling and patting, as you might a dog.

Safe berth

I tied up and went to introduce myself to Chris, the warden of the marina and the owner, with his wife Gill, of one of the most burnished and obviously cherished of the narrowboats. He was as I rather imagined all narrowboat men to be: built on generous lines, genial, tanned and exceedingly relaxed. A surviving streak of radicalism was hinted at by the silver stud in an earlobe. She, too, fitted the type – together they were as fore and aft, breathing decency, welcome, the camaraderie of the waterways.

I told them what I was up to, and Chris produced a camera and took my photograph on the grounds that my visit would make an item for the marina newsletter. They ducked into their cabin, emerged into the sunshine with melodeon and concertina, and proceeded to give me a rollicking rendition of a bouncing tune – an old Northumberland air, they said, called 'Salmon Tails'. I asked them if they did the same for all newcomers. They laughed. No, it was because 'we're think you're doing something really special'. They were heading upriver themselves soon, they said, then up the Trent–Mersey to Stoke. Chris's dad had built one of the bridges over the Caldon Canal and they wanted to take a look.

Times past in Newark

Sure that the *Otter* would be safe with them, I took the land route back into town. I was savouring in anticipation the pleasures to come – bed, sheets, hot water, beer, a plate of meat, potatoes and summer vegetables, a bottle – yes, a whole bottle – of decent red wine. In *Highways and Byways of Nottinghamshire* J. B. Firth dwells at length on the wealth of old inns to be found in Newark's Market Square. There was the Saracen's Head, owned in distant times by the Twentyman family and patronised by Sir Walter Scott, who put it in one of his novels and referred in his diary to the 'remarkably civil and gentlemanly manners' of its landlord. There was the Kingston Arms, where

Byron stayed to oversee the printing of his first verses, and from an upstairs window of which (by then it had been renamed the Clinton Arms) Gladstone was addressing a large and noisy election crowd when he was interrupted by the flinging of a brick that missed his head by inches. There was the Queen's Head, with the Old White Hart opposite, although that had already been converted into a draper's shop by the time Firth's book appeared in 1924.

Weary and smelly, I tramped over the bridge, up Castlegate and into Market Square, where, so Firth had written, 'the tide of activity flows fastest and fullest'. It was almost deserted except for flocks of pigeons and, at the north-western corner, a gaggle of customers who had spilled out of a J. D. Wetherspoon pub called the Sir John Arderne. Arderne is regarded as the first English surgeon, and celebrated for his success in treating a condition known as *Fistula in Ano*, a painful growth between the base of the spine and the arsehole caused by sitting on a horse for long periods in full armour. I was perfectly happy that he should be so honoured. But where was the Clinton Arms, where ninety horses were once stabled? The name was there in big, olde-worlde letters, but underneath it there were shops. Gone, too, were the Saracen's Head and the Queen's Head, while the Old White Hart was now the premises of the Nottingham Building Society.

Much downcast, I tried the Ram in Beastmarket, where George Eliot once stayed, enjoying 'some charming quiet landscapes on the Trent'. Over a pounding jukebox the landlady shouted at me that they were full. I wandered back through Market Square and beyond, and came at length to the Rutland Hotel, where a girl on the desk said they had a free room. I asked her what time they served dinner. She looked at me pityingly and replied that they only 'did food' at lunchtime. She recommended a 'nice Indian' in Stodman Street.

Upstairs I peeled off my clothes and went into the shower room. The shower was broken, the shower head was on the floor and no hot water could be coaxed from it. I washed myself as best I could in the basin and, weak with hunger, set off in search of food. I tried the pizzeria near the bridge, but it was full and the girl on reception said that when it stopped being full it was going to shut. I found the 'nice Indian' and ate a large plate of rice covered with brightly coloured glop and chunks of what might have been chicken, and drank Cobra lager brewed in England. An extractor fan was whirring loudly outside my window when I went to bed, and woke me up soon after 5.00 a.m., although about an hour later it suddenly went off.

Busy river

Despite all these disappointments, I thoroughly enjoyed my saunter around Newark on Monday morning. It is blessed with its splendid cobbled Market Square, several streets lined with delightful old houses, and the glorious parish church of St Mary's with its famously tall, slender spire. I also admired the way the town had taken its river to its heart, whether you call it the Trent — as Newarkers like to — or the Devon, or the Trent Navigation. It has put a fine stone bridge across it, while along the banks stand a sample of the old wharves, mills, tanneries, maltings, cooper-ages and warehouses that drew their life from the water and made the town prosperous. True, none of them has functioned in its original role for decades, and there is almost no commercial traffic any more. But at least Newark shows an affection for its river past, which is more than can be said for Nottingham.

Newark in the 1930s

Then there is its castle, the ruined, empty, magnificent shell of pale walls washed at their feet by the river as they were 350 years ago, when its story came to an end. And what a story it was, stuffed with bravery, dash, daring, defiance, poetry, acts of folly, nobility and betrayal – a story of loyalty to a doomed cause; a much better

Castle

story than anything attached to Nottingham's castle, which was good for hanging boys but not much else. But then Nottingham, as I say, never cared much for its river.

It was typical of Charles I's flair for choosing the wrong option that he should have decided to begin his sacred mission to rescue the English monarchy in Nottingham.[2] The day of the Raising of the Standard was 22 August 1642, a Monday, and the occasion came close to fiasco. The banner was unfurled atop its red pole within the precincts of the castle, and then carried by a score of loyal knights to an eminence behind the outer wall known thereafter as Standard Hill. The King went with it, accompanied by his sons, the Prince of Wales and the Duke of York, and his gallant nephews, Prince Rupert and Prince Maurice, who had dashed over from the Continent to pledge their swords to the cause. Troops of horse and companies of infantry marched behind.

The foot of the pole was thrust with some difficulty into a hole in the rock, whereupon a herald prepared to bellow forth the royal call to arms. At the last moment Charles, typically, decided he wished to change the wording. He called for pen and ink and the company stood around while he subbed the text. At last he was satisfied, and the proclamation rang forth – though rather haltingly, as the herald had difficulty in deciphering some of the corrections. Hats were hurled aloft and the shouts of 'God Save the King' rose with them.

That represented the zenith of Nottingham's enthusiasm for the King. Despite the ceremony being repeated on each of the next three days, a pathetic total of three hundred volunteers came forward. Charles was understandably disgusted, all the more so when he was presented with a petition urging him to return to London and negotiate peace with Parliament. He abandoned Nottingham, where he had received – in Clarendon's words – 'so many mortifications', and decamped to Shrewsbury, where they had some idea of how to make a divinely appointed monarch welcome. In December – following the first major battle of the Civil War, at Edgehill – Colonel John

Hutchinson organised the seizure and defence of Nottingham on behalf of Parliament and with the compliance of most of its people. Thereafter, despite the odd foray up to and even across Trent Bridge, all Royalist efforts to win the city back came to nothing.

Newark did its level best to even the balance. Very broadly speaking, the Trent divided England, with the Royalist strongholds to the north-west and Parliament's power bases to the south and east. Newark commanded the only two bridges over the river downstream from Nottingham. It was also situated at the tip of an arrowhead of Royalist country that separated the King's enemies in Lincolnshire from those in Derbyshire and Yorkshire. It never seriously wavered in its devotion to the Cavalier cause. For that, and its strategic importance, the Roundhead commanders both detested it and longed to possess it.

Colonel Hutchinson had a try in 1643 and came close. The next year the combined forces of Sir John Meldrum, Sir Michael Hubbard and Lord Willoughby had the town surrounded and panting for relief. Indeed, Meldrum was sure that he had struck the mortal blow when he took Muskham Bridge after crossing the Trent on a pontoon. But, like others, he had not reckoned with Prince Rupert and his cavalry. Summoned from Chester, Rupert sent a message to the despairing governor: 'Let the old drum be beaten on the north side early in the morning.' Sure enough, he and his horsemen swept down on Meldrum's army and drove it on to an island in the river, while the Newark garrison sallied forth and retook the bridge. Meldrum surrendered ingloriously and stirring deeds were celebrated in stirring words:

They have won the bridge, those troopers! They will keep it
to the death,
And the foes are drinking hard in the crimson stream
beneath;
And down the gray hillside Rupert's foot is marching in
And echoes high the battle cry, 'For God and for the King!'[3]

That was as good as it got for Newark. Rupert took Meldrum's weapons and clattered off to disaster at Marston Moor. Naseby, in June 1645, spelled the end. For the rest of that summer, the King wandered through England and Wales, chewing over his troubles and various improbable schemes for getting him out of them. He was in Newark briefly in August, and again in October, by which time his remaining supporters were at each other's throats. While they snarled and bickered, Charles did what came most naturally, which was to dither, hoping against hope that Montrose would come south from Scotland to save him. Montrose went the opposite way, and the infighting among the King's men turned more vicious still.

The game was up, and on the night of 3 November, Charles — king in name but little more — left Newark for Oxford, the last place in England where he might be safe. Parliament's forces closed in, but no attempt was made to take Newark, which continued to go about its business that winter. In April 1646, the King — disguised as a servant and having had his love-lock lopped off — was spirited out of Oxford and taken by devious routes to Southwell. There, in the Saracen's Head, he was offered a deal by the Scottish Presbyterians: accept the Covenant and we will put you back on the throne. He agreed, having no intention of honouring the terms, and gave himself up to General Leslie at Kelham, from where he sent orders for Newark to surrender. The King was taken away north, and the defenders of Newark marched out, their heads held high.

They were allowed to keep their swords and their horses, and to depart unmolested. But the castle was another matter. It had caused Cromwell and his commanders too much trouble, and there was always the chance that it might do so again. The order came from London that it was to be 'slighted'. Over the next few weeks everything except the northern gateway and the walls and towers overlooking the river was reduced to rubble. And that is how it has been ever since, a highly picturesque shell of a once impregnable fortress. Now the shell encloses a neat and pleasant municipal garden, with plaques to record some of those brave deeds of old, and hard benches under spreading sycamores on which to ponder the lessons of history, read the paper or drink cans of warm cider.

I did a little pondering and then headed off to buy some bacon at Porter's, the old-fashioned butchers in Market Square. I passed the Ossington Coffee Tavern, an extravaganza of gables, high chimneys and oriel windows presented to the town in the 1880s by Viscountess Ossington, who was moved by 'the earnest desire to promote the cause of temperance'. I wondered vaguely what this high-minded benefactress would have made of its incarnation as Zizzi's Pizzeria.

Market Square itself was fulfilling a twenty-first-century version of its age-old function. It was filled with stalls selling racks of clothes from Oriental sweatshops, stacks of watches, batteries, rolls of plastic sacks, packets of screwdrivers, piles of dishcloths, saucepans, crockery and hideous rugs. A fair crowd was milling around scooping up cheap rubbish by the armful. This fierce appetite for cheap consumables seemed far removed from the Newark of old, a sturdy, useful market town dealing in useful commodities like grain, malt, flour, coal, salt, seedcake, wool, plaster of Paris, saddles and bridles. It had succumbed to the global trash trade. Looking out of the window of Porter's, I found it impossible to imagine stirring events – a Rupert at the head of his thundering cavalry, a Gladstone orating on to the heads of a seething throng – ever happening in Newark again.

There was no one around to serenade me out of the marina when I left that afternoon, and I never did find out if I made it on to the front page of the newsletter. The sun was blazing down again as I rowed back up to the lock in the centre of town. The current –

which I hadn't noticed being with me the day before – was steadily and exhaustingly against me, and by the time the lock-keeper hailed me again I was panting hard, wet with sweat and as red in the face as a ripe Cox's Orange Pippin.

In front of me in the queue for the lock was a spanking burgundy narrowboat, its brass fittings shining like gold and its hanging baskets spilling over with lobelias and pelargoniums. After a short-lived debate with my conscience about the ethics of cadging a lift, I asked if I could cadge a lift. The crew – another man with a big, jolly, bearded face and extensive stomach, and a jolly wife who would undoubtedly have had a beard if she'd been a bloke – said they'd be delighted. They towed me and the *Otter* back to the point where *Power In Trust* jutted at the sky. I asked them where they were heading, and they just looked at each other and shrugged. Wherever the water and the fancy took them was the answer.

I parked the *Otter* and went to prospect Averham Weir. Like Beeston, it looked and sounded more daunting than it was. It is a rather uncouth obstacle made of large boulders and lumps of

White water Averham

masonry spread in a ragged line across the head of the unnavigable channel, at an acute angle to the Newark-bound diversion.
The current down the left bank, immediately below *Power In Trust*,
was fierce and I would not have fancied it. But there is a big island
in the middle, thick with white willows, with a secondary channel
beyond. This, for all the foam and white water, was pretty
innocuous. I dragged the *Otter* over the sill and coaxed her down
a little tumbling cataract into a pool with a gentle eddy at the near
side fringed by a grey beach of crushed snail and freshwater mussel
shells. Grass sloped down to the beach, and a big weeping willow
spread shade over it.

I made my last, best, camp there. I put up the tent, and then
lay on the grass, my head on a cushion, my bare feet resting on
the moist, rasping sand. I read *The Mill on the Floss* for a while, then
looked up at the waving willow fronds and the sky, and across the
bubbling water to the island. It screened me from the footpath
than ran along the left bank. No one could know I was there. I
was hidden, kept company by the trees and the river, soothed by
the music and dance of the water.

The current, having fragmented itself to get over and through
the boulders of the weir, reassembled itself in my pool. While the
eddy curved along the beach, the main flow arced towards the
island, then back towards the near bank, pulling at the trailing
branches of the next willow down. It was shallow enough in places
for me to see the bottom, bronze gravel winking in shafts of
sunlight. The surface seethed and frothed, broken into competing
bands. It looked irresistibly fishy.

My friend at the Environment Agency had told me that the
fishing at Averham Weir was good. Very good. But forget it, he
said. Don't be tempted. The water was controlled by a club called
the Nottingham Piscatorials. They were long-established,
extremely careful in selecting members, sticklers for their extensive rule-book. 'They always prosecute,' he warned me. 'No

exceptions.' Sure enough, I saw the signs when I explored down-stream from my camp: 'Nottingham Piscatorials. Private Fishing. Members only. No day tickets.' However, it was clear from the barrier of virgin vegetation along my side that none of these piscatorials ever came this way. The anglers' pitches were along the far side, and I and my pool would be invisible. It was virgin water, and the temptation was overwhelming. I was overwhelmed.

While the sun sank upriver, I ate the last of my baked beans with the bacon I had bought that morning, and drank the last of my last bottle of Rioja. Golden light slanted through the willows, irradiating the honeydew dropping noiselessly from the foliage, so that I seemed to be sitting in a shower of golden rain. At last the sun descended behind the trees on the island. The water darkened, dippers flitted between the rocks, sedges hatched and scuttered for the margins.

I climbed as quietly as I could into the *Otter* and guided her across to an outcrop at the head of the pool where a miniature willow cast a shadow on the water. I tied her to the willow so that she swung gently beside the deeper water. I had half a tin of pork luncheon meat left. I cut up a handful of little cubes and scattered them on the water, to give the fish a taste. I impaled a bigger cube on the hook, and swung the bait and a weight out so that they plopped together on the edge of the current.

I knew I was going to get a fish. Something in the way the water rippled and twisted and turned back on itself; something in the way the threads accelerated and slowed, merged with each other and then separated; something in the look of the water, its amber colour; something in the way the willow's branches dipped to meet it – all these activated the instinct within me. I could feel fish almost as if I could see them. I knew that with the cool of night after such heat, they must feed.

I held the rod across my knees, angled downstream, the line caught between the thumb and forefinger of my left hand. I could feel the nylon trembling with the river's energy. Occasionally there was a quiver as the bait shifted on the gravel. Then there was a bang, the tip of the rod pulled over, and after a brief struggle I netted a fat, handsome chub about the same size as the one I'd caught at Burton.

It was a decent start, but not what I was after. I put on another greasy chunk of meat and cast again, further down, about a yard out from the trailing willow branches. The dusk gathered. I smoked a cigarette, vaguely aware of the smoke mingling into the warm air, but with every conscious faculty intent on the hidden water world. Another bang, again the rod went over. This time the resistance was much sterner. The fish bored for the willow roots, then across, using the current. It forged its way up the far side, filling me with fears for sunken logs and jagged rocks. I saw the gleam of its flank through the water, bronze like some Roman serving vessel being retrieved from a resting place. I hauled it to the surface and slipped my little net beneath it. It was a barbel, about seven pounds. A very special fish, and well worth risking a period of imprisonment.

I stopped then. I didn't want any more. I am as happy as I could be, I thought, as I struggled into my sleeping bag for the last time.

Chapter 18

The Church by the Stream

At Averham the river briefly threw off the tyranny of diggers and dredgers and hydraulic engineers and recaptured the ebullience and energy of its Staffordshire adolescence. It raced over shallows, surged through the pools where the gravel shelved down, circled around on itself in sweeping eddies, dashed back and forth like a dog let off the lead and thrown a stick to chase.

Looking for breakfast

All I had to do was to paddle out from my campsite and let the river take over. We hurried past the willow where the barbel took, and past leaning alders. I dipped the paddle now and then to keep us to the front and away from mudbanks and gravel bars. There was a cobalt flash, too quick for the eye: a kingfisher patrolling its territory. A heron stood in six inches of water just out from a bed of reeds, a pale, motionless hunter.

Below the island the two channels came together in a vee, ran smoothly down to the railway bridge, then quickened through the arches. We scooted through the middle one, from dazzling sunlight to deep shade to dazzling sunlight again. To the left was a wide, flat meadow. Ahead, nestled among high beeches and dark yews, I caught sight of the top of the square tower of Averham church. Of all the Trentside churches, it is the closest to the water, and perhaps also the sweetest.

Averham Church

'They chose well who chose this site and they built a church worthy of the setting,' wrote Firth in *Highways and Byways*, a judgement that has lasted better than his description of Newark's Market Square. The original 'they', in Norman times, was one Gislebert Tyson, from whom the manor of Aygrum, as it was called then, was passed to one Robert le Sauvage.

Thoroton relates that this Robert came to owe a 'great sum' to a Jewish moneylender named Aaron, and turned for help to a neighbour, Robert of Lexington. It so happened that this Lexington had a brother, John, who was a priest in Lincoln when the body of a boy was allegedly recovered from the well in the garden of a prominent Jew named Copin. Copin was accused of abducting him, torturing him and crucifying him in mockery of Jesus Christ. The priest obtained a confession on the promise that Copin's life would be spared; but a month later Henry III came to Lincoln and ordered that Copin be dragged through the city attached to the tail of a horse and then hanged. Eighteen other Jews were executed and seventy more were flung into prison. No evidence to sustain any of the charges was ever produced, and it's almost certain that the whole episode was orchestrated as a pretext to seize Jewish property, cancel debts and demand ransoms for those in jail.

Still, it turned out well enough for the priest's brother, Robert, who got the manor of Aygrum from le Sauvage. In time he left it to a nephew, Roland of Sutton (Sutton being a few miles down the Trent). And the Suttons remained the 'they' in Averham and the neighbouring manor of Kelham for the next 650 years, until they got too full of themselves and lost the lot.

Part of their long story is told by the silent figures and tablets inside the church. Sir William Sutton, deceased 1611, lies there encased in his alabaster tomb, beside his wife Susan. The inscription celebrates their union and its blessings:

> Thrice nine years liv'd he with his lady faire,
> A lovely, noble and like vertuous payer:
> Their generous offspring (parents joy of heart),
> Eight of each sex: of each an equal part
> Ushered to heaven their father, and the other
> Remained behind him to attend the mother.

On the opposite wall is a casket ornamented with appropriate shields of arms, said to contain the heart of one of the eight who stayed behind to look after the mother. This was another Robert, who was made Lord Lexington by Charles I for loyal service in the

Sutton at rest

various sieges of Newark, and expressed in his will the wish 'to be buried near my last wife in the Church of Aram among my ancestors; and I hope my present wife, if no new affection make her forgett me, will come and lie by me there and we shall rise together at that great day of the Lord.'

In time the estates beside the Trent passed on the distaff side to the Duke of Rutland's son, Lord Robert Manners, who added the Sutton to his name. The Manners-Suttons went from strength to strength. One, having done his time as Rector of Averham, advanced through the clerical ranks to become a worthy Archbishop of Canterbury. Another, the Archbishop's younger brother, went in for the law and became Lord Chancellor of Ireland, where Daniel O'Connell encountered him and judged that 'he was the most sensible-looking man talking nonsense I ever saw'. One of the Archbishop's sons, Charles, had his career in politics cut short by a fatal apoplectic seizure while travelling on a Great Western night-mail train. 'He appeared to have been in perfect health as far as Slough,' one newspaper reported, 'and kept up a lively and most agreeable conversation. Soon after leaving Slough he was seized by a fit.'

Various Suttons and Manners-Suttons had the living at Averham, one of whom built the large and handsome rectory next to the churchyard. Eventually, however, there was no one left in the family who wanted it or was judged worthy of it. It was discreetly advertised and sold for the enormous sum of £16,000 to a hugely wealthy Liverpool lead smelting magnate, Joseph Walker, who installed his younger son as rector.[1] This Walker, also Joseph, lasted forty-five years and was succeeded by his son, Cyril. Cyril had strong theatrical leanings as well as plenty of money, and built a theatre in the garden where he staged plays and light opera, including *Aladdin*, in which the young Donald Wolfit played the part of the donkey.

There are no Walkers in Averham any more (the Reverend Cyril lasted until 1942). But the theatre is still a theatre, and the church is still there, and the rectory, which is owned (or was when I was there) by a London property developer who had enclosed the gardens within high fences and installed guard dogs to keep the riffraff out.

I set these beasts baying and slavering shortly before eight o'clock that morning. I'd tied the *Otter* up in a bay next to the graveyard, intending an early appointment with Sir William Sutton and his fertile wife. The church was locked, and a card on the door said a key could be obtained from a house up the lane. I did wonder about the hour, but I needn't have, as Canon David Keene greeted me warmly in dressing gown and slippers and handed over the key without a murmur of complaint. I spent a pleasant half-hour at St Michael's and All Angels, admiring the Sutton memorials, the heraldry, the gargoyles, corkels and quatrefoils, and the dark pews and panelling provided by Cyril Walker.

By the time I returned with the key, Canon Keene was dressed in short-sleeved shirt, shorts and sandals and was evidently in the mood for some chat. He had also brushed his hair, although nothing could subdue the exuberance of his leaping white eyebrows. He was a clergyman of an old-fashioned and delightful kind who had spent much of his working life in Nottinghamshire – at Radcliffe-on-Trent and subsequently at Southwell. He had a vigorous way of talking, pushed along by the frequent bark of 'd'you see, d'you see?' accompanied by a jump of the eyebrows and a keen look to make sure I was paying attention.

He'd known Averham from long ago, when the rectory was maintained as a Church of England retreat. As a young man, he had stayed there to pray and study. Others were dispatched there to recover from alcoholic or other forms of excess, or to be kept out of the way until some scandal or other had died down. I told him about the *Otter*, and he recalled his punting days at Cambridge. The Cambridge style was to wield the pole from a platform at the rear, and he was contemptuous of the Oxford way of roaming up and down the boat. I hardly dared tell him that I was paddling and rowing, so I asked him about Southwell. He said it was 'too cluttered' for his taste now. He told me a story about his predecessor being asked if he would mind not having the Prebendal house to which he was entitled and replying: 'I couldn't care less where I live as long as there's enough room for my train set.'

I enjoyed talking to the canon enormously, and I was still smiling as I unhitched the *Otter* and left Averham behind. Hardly had we done so than my attention was taken by the rooftops and mighty chimneys of an outlandish building: Kelham Hall, the folly that caused the downfall of the Manners-Suttons.

* * *

Kelham Hall 1829

239

In November 1857 John Henry Manners-Sutton and his wife were in the English Club in Naples when they got news that their fine but unremarkable home on the banks of the Trent had been destroyed by fire. The blaze broke out just as a major programme of modifications and modernisation was being completed, which had involved fronting the building in stone, adding towers and a balcony to the top floor, and building a new kitchen wing. In its account of the fire, the *Newark Advertiser* took a dim view of some of the craftsmen employed on the project: 'We cannot too strongly censure the stone masons who were cruel enough, when asked to give a helping hand, with a contemptuous laugh and a shake of the head to decline.'

Scott's drawing

The most recent works had been supervised by the great champion of the Gothic revival, George Gilbert Scott, and Mr and Mrs Manners-Sutton promptly commissioned him to build an entirely new house. In the words of the architectural historian Mark Girouard, Scott 'was now presented with an empty site, a compliant patron and what seemed a long purse'.[2] During their travels on the Continent, the compliant patron and his lady had been much impressed by the Alhambra, and they told Scott that they wanted 'something Spanish'. 'The resulting phoenix of red brick,' wrote Girouard, 'that emerged from the ashes still surprises travellers who cross the River Trent at Kelham; and its amazing silhouette enriches the view for miles around.'

Girouard and Pevsner are both pretty sniffy about the vast, forbidding pile, with its clutter of gables, towers, balustrades, dormers, high chimneys and almost infinite diversity of windows; and its fabulously sumptuous interiors, bristling with vaulting, gilded capitals and mouldings, marble piers, hooded chimney pieces and the like. Leonard Jacks, in *The Great Houses of Nottinghamshire and the County Families*, was more easily pleased.[3] His account is weighed almost to the ground by superlatives: the wondrous decorations, the exquisite taste, the delicacy of execution, the delicious odour of the cedar panelling in the Small Drawing Room, 'the most delicious marble of that rare and richly marked creamy variety', the elaborately wrought mouldings, the pillars of the Principal Drawing Room 'of Australian marble, dark and wavy'. Can he really have meant marble from Australia?

Understandably somewhat dazed by the splendour, Mr Jacks eventually finds himself in Mr Manners-Sutton's business room, where 'the table is strewn with letters, papers and books of reference, for Mr Manners-Sutton is a thorough man of business'. But the poor man needed a miracle rather than a head for business to rescue him from the mess he was in. The equation was simple. The estate brought in £11,500 a year, and the cost of the house was £80,000 and still rising. Even the fawning Mr Jacks could not fail to notice that 'one or two of the rooms are yet in an unfinished state'. In fact the conservatory never got off the ground, the clock tower had no clock, the chapel was left undone, and throughout the house there were empty sockets still awaiting their columns.

The last of the line

Mr Manners-Sutton was ruined. 'One can only guess at the reasons for his extravagance,' Mark Girouard commented solemnly. But maybe there wasn't a reason, in that severe, logical sense. His head was turned, filled with apprehensions of wonder and beauty. He did not wish to be another Manners-Sutton in the long line of Suttons and Manners-Suttons, another Nottinghamshire squire like all the others. There was nothing much he could do about the situation – the river, the monotony of fields, the straggling hedgerows and copses and coveys, the town across the way with its racket of manufacturing and trade. But he wished to look out over the familiar scene, not from a solid squire's solid residence, but from a palace, a palace decked out in the manner befitting a Renaissance prince.

In 1890 the last of the Manners-Suttons of Kelham was paralysed by a stroke. His only son had long since been shut away in a home on account of 'morbid shyness'. Manners-Sutton died in 1898, whereupon the mortgage was foreclosed, and Scott's grandiose monument came into the hands of the Society of the Sacred Mission, an offshoot of the Church of England. It is now the headquarters of Newark and Sherwood District Council, and is kept pretty spick and span, although various external additions and alterations have given it the appropriate municipal feel. It's odd to think of the plush chambers where Mrs Manners-Sutton would have reclined fanning herself over the latest number of *Blackwood's Magazine* or *Punch* being used to dispatch council tax demands, process planning applications and examine ways of meeting recycling targets.

Kelham Hall

I had only a few moments to take in the monstrous mag-nificence of Kelham before the current took us beneath its stone and brick bridge and on. More bridges followed: one old one, forming part of a great causeway upholding the Great North Road above the floodplain; then a metal railway bridge. Another loomed, a concrete single span of the functional variety for the A1. But I did not go under it. On the left, just past the railway bridge and opposite the old village of Winthorpe and its no-so-old sewage works, was a bay with a sandy beach. I grounded the *Otter* gently. It was time to go our separate ways.

Otter at the end

Chapter 19

Redeeming Mr Sweetapple

Stanley would have gone on in the boat. So would Thesiger. And Mungo Park, and Sir Ranulph Fiennes. But I funked it. I had my reasons, but a braver traveller than me would have ignored them. I rather regret it now, as I've always regretted not being more intrepid. But there you are.

Three or four miles below Crankley Point – where the two arms of the Trent are united, and where the *Trent Otter* and I were disunited – is Cromwell. It is at the top of the Trent's seventy-mile tidal stretch, and there is an enormous weir and a lock big enough to take oil carriers. Downstream from Cromwell different rules apply.

On the night of 28 September 1975 eleven members of the Parachute Squadron of the Royal Engineers were taking part in a night exercise called Expedition Trent Chase. The river was in flood after days of rain. Towards midnight their inflatable assault craft came downstream towards Cromwell Weir, which in those days did not have a boom across the top. As it did so a violent thunderstorm broke. The lights above the lock blew and the emergency generator failed. The lookout posted on the bank missed the boat as it passed. In the inky blackness, the roar of the ten-foot sheer drop was lost in the booming thunder and driving rain. They went over and were sucked in. All but one of the party were drowned.

There is a plaque at the lock commemorating this awful event.

I went to look at it. I looked at the lock, too, and imagined the *Otter* lost in the bottom of it. I looked at the weir, which even at low water was an alarming sight. I talked to the lock-keeper who pointed at the swirling chaos of water. 'If you get into that,' he said, 'you don't get out.' He explained to me why, something about the steepness of the drop and the roll of the water back on itself, but I wasn't really listening.

I read with close attention the navigational notes included in *Nicholson's Guide* for the benefit of those venturing on to the tidal Trent: 'A suitable boat is essential. Proper navigational lights and safety equipment (including an anchor and cable) is [sic] also compulsory . . . boats should be aware of shoals at low water and should avoid the inside of bends . . . the river banks are unsuitable for mooring boats . . . there are few wharves . . . navigators should plan their trip with an eye to the tide tables . . .'

There was food for thought here. Was the *Otter* a suitable boat? Where did I find navigational lights and how did I fit them? Did a lump of stone and ten yards of rope constitute an anchor and cable? Did I qualify as a navigator? How was I supposed to familiarise myself with the warning sounds and signals of commercial craft?

The Aegir on the Trent

Aegir again

My imagination was working. How would I manage at night? I would have to sleep on the boat. But what if the tide came in when I was asleep and floated me off downstream? Or upstream to Cromwell Weir? What if it went out and I was stranded halfway up the bank? What if I was confronted by the Aegir, the Trent version of the Severn Bore, which sweeps a seven-foot wave upstream at the speed of a galloping horse? What about the vast gravel barges that churn up and down from Hull and Goole spreading great rolling wakes that could quite easily overwhelm an unsuitable boat like – say – a plywood punt? And what if the skipper of the barge didn't see me (him being drunk) and I didn't see him (me rowing, with my back to him)? What if I lost an oar? Or a rowlock?

I tried to rationalise my fears into strategic thinking. I asked myself what I would gain by staying on the water. The answer was obvious. This was a river journey and I had got so far and I should keep going. But what might I lose, apart from my life? This was a telling point. It is a fact that, from Cromwell down, the river is confined between flood banks high enough to block the view from river level of anything beyond the river. I saw myself trapped, imprisoned within river walls, fearful of disaster at every turn.

I talked to my friend at the Environment Agency who had urged me over the weirs at Beeston and Averham. He advised me not to go beyond Cromwell. I said, why not? Canoeists do it. He said canoes were different from punts and canoeists were mad anyway. He spoke solemnly about the scale of the river as it neared the sea, its unforgiving character, the hazards of the tides and the barges. Further down there would be not just barges but ships. Ships! Then he told me about the Royal Engineers and Cromwell Weir.

That was enough. I got on my bike.

My wife came with her parents to meet me at the *Otter*'s last resting place, and to deliver my bicycle. She took the punt away on a trailer and I cycled into Newark to get two new tyres and then out again along the right bank. For the first time in a long while I was studying the map, not out of interest as to where I might have got to, but from the urgent necessity of working out my way. It took me down a lane to the hamlet of Holme.

Minute and mute, Holme communicated the sense that it had never quite recovered from being moved from one side of the Trent to the other. From the earliest times it was the neighbour of North Muskham on the west bank and intimately associated with it for administrative purposes. Then, sometime in the sixteenth century, the river changed its course and separated the two communities. Through no fault or desire of its own, Holme was dumped on the east side with nowhere and no one to keep it company. That is how it is now. No settlement between Newark and Gainsborough, twenty miles to the north, is closer to the river or more cut off from everywhere else.

Looking at the river now, sliding unobtrusively along with the villages either side protected behind the embankments, it's difficult to imagine the extent to which people used to be at the mercy of its caprices before the engineers went to work.[1] Ten thousand years ago, this stretch of the floodplain was covered with vast layers of gravel as the ice sheets retreated. The river split and split again with each flood, creating a web of channels through the gravel. The wide, flat valley was a highly unstable environment, and living on it was a precarious business. The river followed its whims, cutting new courses, shifting gravel bars, switching its weight from one channel to another almost as the mood took it. It would lengthen itself, extending its meanders into ever more exaggerated loops, then suddenly force its way from one bend to the next.

Here, the main flow for centuries, probably thousands of years, was to the east of Holme, possibly along what is now a negligible watercourse called The Fleet. But there were other fault lines in the landscape available to the river as it sought to migrate west. One winter, perhaps between 1550 and 1575, a flood forced its way down the narrow dimple between Muskham and Holme and split them asunder.

A great deal of the old history of the Trent is about floods. The end of the Ice Age had left no natural watershed between the river's course north to the Humber, and its pre-glacial path east to the Wash. It was forever trying to elbow its way back; and although it could not succeed, it made plenty of mischief. There were incursions most winters, and major invasions at intervals: 1346, 1683, 1770, 1814, 1839, 1852, 1875, 1887 are just some of the flood years recorded. In February 1795 fifteen inches of lying snow melted almost over night, setting off a tremendous downstream surge. Outer flood defences between Spalford, Girton and Collingham to the east were breached, and 20,000 acres of land extending to Lincoln became a lake. The water rose to a height of four and a half feet against the wall of North Collingham churchyard. This and later flood marks were recorded on the wall of a house in Girton, with the little ditty: 'When this you see/Pray think of me.'

* * *

Holme Church

I doubt if Holme was a bustling commercial centre even before its banishment to the other side of the river. But at least it had its big house, and had had its resident big noise. He was John Barton, who made a fortune in Tudor times from the wool trade with Calais and showed his appreciation by having engraved into one of his windows the lines: 'I thank God and ever shall/It is the shepe hath payed for all.' Barton lies beneath the east arch of the little church he endowed, dressed in his wool merchant's finery. His wife, Isabella, is beside him, her feet resting on a lapdog, and very peaceful the pair of them look. Over the porch of the church is a tiny room known as Nan Scott's Chamber, where, so the story goes, a village woman by that name shut herself away to escape the Plague. When she came out, so the story continues, she found only one other survivor and was so traumatised that she retreated to her chamber and was never seen again. The story, I'm sorry to say, has now been reclassified as legend; probably the work of an inventive freelance journalist.

I sniffed around the church and the few houses (John Barton's fine mansion disappeared a long time ago), then took a path which, on the map, promised to take me along the east bank to Cromwell and beyond. Having heaved my bicycle over a succession of stiles, pedalling in between, I reached Cromwell, only to find the path blocked by a metal fence festooned with barbed wire from which hung a notice stating, without explanation or apology, that it was closed.

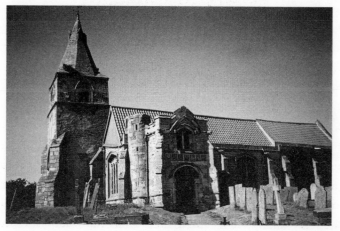

and again

There was nothing for it but to return to Holme. The way back seemed longer and the stiles mysteriously higher, and by the time I left Holme for the second time I had had quite enough of it. The afternoon was wearing on into evening, and I was beginning to think about the night and where I might spend it. Having said farewell to my tent, sleeping bag and cooking gear along with the *Otter*, camping was out, and my ambition extended beyond hedgerow or bus shelter. Reviewing my experiences in Newark, I decided I wanted a hot shower as much as I had ever wanted anything in my life.

I took the view that there was no need for me to try to stay with every twist and turn in the river's course. In the boat there was no alternative, but the bicycle offered choices. The twists and turns all looked very much the same, and the way along the top of the flood bank was overgrown with long grass and extremely arduous to pedal for any length of time. From Holme I sought and found the boring, noisy main road north. Collingham, Girton, South and North Clifton passed by without making any impression on me. Just out of North Clifton I stopped being in Nottinghamshire for the first time in quite a while, but Lincolnshire looked just the same. At about half past six I turned into the village of Newton-on-Trent, which wasn't on the Trent at all but about a mile away. However, it did have a pub, the White Hart. They did B & B, they had a free room, they did dinner, the shower worked, the bed was soft and inviting.

Thunder muttered and growled at intervals during the night. Next morning I found that, under cover of darkness, a moist, chilly blanket of misty murk had crept across from the North Sea. The landscape – flat fields spiked with pylons and tied up with looping cables – had been drained of colour. The landlady presented me with a life-enhancing fried breakfast. As she fussed about the dining room, she told me that she never got to bed before midnight and was generally up at six. It was, she said, the only way to make a go of it. I asked her if she got a lie-in on Sundays and she laughed. Twenty-five anglers came every Sunday of the coarse fishing season, from mid-June onwards, paying a fiver a head. I felt rather humbled by her energy and cheerfulness. And I was glad, too, that the fishermen still came.

One of the things I was seeking on this journey was a connection with the moving water, to be on it and of it, to have it flowing around and through me, if that doesn't sound too pretentious. This was comparatively easy in the upper reaches, but had become progressively more difficult as I got further downstream on the Navigation. Now that I had left it, it was harder than ever. Partly, of course, this was because I had left it; but it was also because of it, the river. It was so thoroughly tamed, controlled, directed, devitalised that it seemed to have surrendered much of its essential riverine quality. Something similar had happened to its landscape. It had been reduced to a site for agricultural production and power stations, largely devoid of people, staffed instead by triangular steel giants combining the duties of servicing the cooling towers and keeping watch on the fields.

It had not always been thus. It is only over the past hundred years or so that we have achieved this degree of dehumanised control. Go back beyond that period, and you find a different land and a different river.

For my friend J. W. Martin, the 'Trent Otter' – I think of him as a friend even though he died before the First World War – the golden age was the 1860s and 1870s. Anglers are in a better position than most to commune with water and the life within it, and Martin got to know the Trent and grew to love it when it was still a real river. In a book called *My Fishing Days and Fishing Ways*, he did honour with a simple, wonderful vividness to the water and his fellow fishermen.

They came from distant, smoky Sheffield, those 'horny-handed grinders and cutlers', by train to Torksey to the north or Tuxford to the east, then walked the six miles or so to the river to be in position 'as the first grey streaks of dawn appeared on the horizon'. Some arrived that morning, others would have bedded down the previous evening at the Bridge Inn at Dunham (just across the river from Newton) 'kept by very old friends of mine, ex-Sergeant Dixon and his wife . . . a kindly village couple of whom I have nothing but kindly remembrances'.

Upstream from Dunham the visiting angler had miles of river at his disposal. There was a 'long, peculiar hill' at Dunham, dropping down to a deep hole known as Dunham Dubbs, the haunt of 'huge quantities of very large bream'. During the summer, Martin's friend Charlie Hudson – the local professional fisherman, long since gathered to join 'the great majority' – kept his houseboat there, tipping thousands upon thousands of juicy worms into the hole to keep the bream happy.

Choice spots abounded in those palmy days. There was a line of bushes on the Dunham side with 'a capital stream running by' under whose 'roots and fastnesses I have had more than one splendid bag of chub'. Big barbel shoaled under the bridge itself, while above the Dubbs was a series of barbel swims 'of the very first rank'. Further upstream, around Collingham, Carlton, Holme and Winthorpe, the chub came into its own. Martin records that these waters were 'much patronized by that famous Nottingham angler William Bailey and his equally famous friend George Holland . . . The chub that swarmed down all those reaches . . . were something to be remembered; it was no uncommon occurrence for those two fishermen to be out for a week living on the proceeds of the sale of their fish to the neighbouring villages.'

Even then (Martin's book was first published in 1906), he was sighing over the departed glory, lamenting that 'the constant dredging of the channel and the drawing away of the water from the banks and bushes has spoiled the fishing without any doubt'. But not all the news is bad. As recently as August 1905, he receives a report that Councillor David Salter caught ninety-seven barbel in a day on the Holme water. And he is heartened to get 'a long and interesting letter from a Lincolnshire coroner who is well-known to Grimsby anglers under the nom-de-plume of "the Lincolnshire Clodhopper"', reporting that the railway has been extended to Fledborough, where visitors will find a welcome at the Brownlow Arms.

But Martin, himself not that far from joining the great majority, is happiest dwelling among his memories:

I love the old Trent and its sparkling shallows. I lived long within sound of its roaring weirs and its gurgling eddies, and know it better than a man knows his most intimate friend. Those deeps and streams would talk to me; the nodding willows would beckon me on; and I could talk back to them in their own language. The very sandpipers even, that flitted from sandbank to sandbank, would show me where the fish swam thickest. I have slept by the side of it with nothing but the sky for a coverlet; have seen it in all its varying moods; have been lulled to sleep on its banks by the magnificent midnight song of the mock nightingale; have woke up and seen nothing but a wide, rolling sea of summer mist . . . Ah yes, twenty years on one river is not wiped from memory's slate as a child's forgotten drawing; but it lives and will live until that unseen but relentless hand beckons us across the gloomy borderland.

If they could see it now, the 'Trent Otter' and the 'Lincolnshire Clodhopper', I fear their first inclination would be to throw themselves into it rather than fish it. All that diversity has been erased — the sparkling shallows, gurgling eddies and capital streams, the grand deeps and swims. The Trent has now one face and one character: steady, placid, ponderous, slightly dull. The consolation, and one that the old-timers would have appreciated, is that the anglers still come because there is reason for them to do so. The barbel and chub are still here. The water, for all its brown, soupy look, is clean and full of life.

Fledborough Church

* * *

The idea that Fledborough should ever have been considered worthy of having its own railway station seems extraordinary now. The sign directing you there from the main road used to read 'Fledborough and no further', and it still has an apologetic air about it, as if to suggest that it really wouldn't be worth your trouble. The lane ends at a fence, with the church on the left and a meadow ahead, beyond which is the river. Behind the church is what used to be the rectory. Opposite is a ramshackle farm, squeezed from behind and the sides by the spectral ruins of High Marnham Power Station. In its day this was the biggest coal-fired generator of electricity in Europe, producing enough to light and heat Nottingham five times over. But its day was over years ago. Now it is no more than another uncompromising statement of indifference to landscape, all the more brutal for its silent uselessness.

High Marnham power station

A note on the church door, faded and curled with age, said a key could be obtained from the farm. I left my bicycle against the wall and walked up the drive and through a yard littered with defunct machines and patrolled by a pack of lively dogs. The roof of the house was sagging and the paintwork on the window frames was cracked and peeling. Behind it loomed the cooling towers, stained with grime. The rest of the plant had either been demolished and removed, or was rotting and rusting away. But the towers looked as if they would, like the pyramids, last forever.

A pink-cheeked girl in mud-spattered jeans, gumboots and an enormous sweater broke off from attending to an elderly tractor to get me the key. The church was simple and lovely: a square Norman tower, and a fourteenth-century whitewashed nave with bays and pillars and fragments of old glass. In the north aisle is the stone figure of Dame Clemence de Lisieux, the widow of one of the lords of the manor, her head cushioned, her feet – like those of Mrs Barton in Holme – resting on a dog. But it was not the Dame who had brought me to Fledborough, but the curious story of the man who was rector here between 1712 and 1757, the Reverend William Sweetapple.

Mr Sweetapple has been treated roughly and unfairly by posterity. Every guide who condescends to mention Fledborough at all has a humorous swipe at him and his allegedly easy-going attitude to the exchange of marital vows. His reputation as 'the rascally Mr Sweetapple' sprang from an examination into Nottinghamshire marriage registers conducted long, long afterwards, which revealed that between 1728 and 1753 he conducted 493 weddings, thereby uniting in matrimony a number of young people fifty times greater than the entire population of his parish.

The disclosure took on a life of its own. 'Gradually it became known,' wrote the Archdeacon of Newark, the Venerable Frank West, 'that in this snug little hamlet on the banks of the Trent, an ideal spot for a clandestine marriage, there was an obliging parson who would grant a licence without asking questions one moment, and perform the marriage ceremony in the church at the bottom of his garden ten minutes later.'[2] The charge was recycled – for instance, by Peter Lord, who, in his book about the Trent, alleged that the Rector had granted licences 'with an abandon never recorded before nor equalled since, attracting to his remote village church runaway couples from all over the country.'

The truth is more prosaic.[3] It was uncovered by an expert on marriage in the eighteenth century, the late Doctor Brian Outhwaite, who went to considerable trouble to find out who these 'runaway couples' were, and where they came from. His findings make it clear that the case against Mr Sweetapple is entirely false. In every instance at least one of the partners – and in most cases both – came from within fifteen miles of Fledborough. The grooms were farmers, agricultural workers, servants, butchers, carpenters, wheelwrights, tailors, drapers, mariners, ferrymen – in Dr Outhwaite's well-chosen words 'a broad band of middling folk.' The charges for licences could not possibly have represented the 'lucrative sideline' claimed by another Trent historian, Richard Stone, and the worst that could be said of Mr Sweetapple was that he took a relaxed attitude to regulations that, at the time, were laxly enforced. People probably came to him because they wanted a quiet wedding. Then, as now, there can have been few places as quiet as Fledborough.

I sat on the organ stool with Mr Sweetapple's tomb beneath me. It struck me that part of his image problem was his comical name, and that if he'd been called West or East or Brown or Smith, he would have been much less likely to have inspired this enduring fiction of shady dealing. On the way out I stopped in front of a white alabaster monument placed there by one of Mr Sweetapple's successors, the Reverend Augustus Fitzgerald. It commemorated Sarah Anne Fitzgerald 'who died on the second anniversary of her wedding day, February 7th 1841, aged 21 . . . and at the same time and in the same grave was placed the lifeless corpse of Sarah Anne the infant daughter of the above. She survived her sainted mother but four days.'

I returned the big iron key to the girl in the enormous sweater and took the path along the flood bank back to Dunham. There were fishermen regularly spaced along the far bank, but they were too far away for me to ask them about the great bream hole of Dunhams Dubbs. I did meet an angler in the shadow of the toll bridge. He was stretched out on a reclining seat with two rods sticking out in front of him. He'd been there since dawn but hadn't had a bite. I marvelled at his patience. 'Yeah, I suppose it's a bit boring,' he said smiling. 'But it's me day off and it's better than being at home.'

Chapter 20

The Mark of Cornelius Vermuyden

The names of our rivers are satisfyingly old, most of them so old that scholars still argue courteously over their origins. Very many are Celtic, some Saxon or Scandinavian, a few even French. Their musicality is one of the delights of geography. Evenlode, Derwent, Duddon, Medway, Nadder, Wylye, Otter, Eden, Eamont, Severn, Thames, Windrush, Loddon, Camel, Fowey, Tamar, Torridge, Annan, Oykel, Naver – all slide from the tongue, quietly appealing for the chance to be taken into poetry or song. The monosyllables – Swale, Blyth, Dove, Dart, Pang, Mole, Colne, Wey, Wharfe, Tyne, Lune, Esk, Ouse, Dee, Tay, Spey, Kent, Nene, Wye, Exe, Lea – are perhaps a shade more prosaic. But they, too, invite melody and rhyme.

Of course the harmonies are no more than linguistic serendipity. They sprang from meanings that were commonplace and endlessly repeated: simple people pointing at something that impinged insistently on their simple lives – the local river – and grunting out a sound of recognition. We have a host of Avons – the Hampshire Avon (which begins as the Wiltshire Avon), the Gloucestershire Avon, the Bristol Avon, Shakespeare's Avon, a little Avon in the New Forest and others – and the sound is smooth and charming. But all it means is *river*, and is the same as Afon in Welsh and Abhann in Irish (the River Bann).

Walton and his contemporaries were understandably clueless about where these names came from. They put Thame and Isis together to make Thames and wrote poetry about the nuptials. In the case of the Trent, Walton guessed it might have come from Trentham (it's the other way round), and followed the accepted line that it must have something to do with the French for thirty. Michael Drayton summoned the river itself to discuss the matter:

What should I care at all, from what my name I take,
That thirty doth import, that thirty rivers make;
My greatness what it is, or thirty abbeys great,
That on my fruitful banks in times formerly did seat;
Or thirty kinds of fish that in my streams do live,
To me this name of Trent did from that number give.[1]

Ha ha. We chuckle at such simple-mindedness. How could it be French when it was named long before the French ever arrived? What dunces!

Yet when I looked a little way into the subject of river toponomy, I found it a surprisingly dark place four hundred years later. The supreme authority, quoted in every discussion of the matter, is a book called *English River Names* which was published eighty years ago and written by Eilert Ekwall, a Swedish professor from the University of Lund.[2] Far be it from me to question Professor Ekwall's competence. I know almost nothing about the matter; he, as a glance at any page of the great work shows, knew a vast and intimidating amount. Yet quite often, when it comes down to the nitty-gritty, Ekwall is surprisingly hypothetical, not to say opaque.

Take his discussion of the etymology of Thames. It ranges far and wide, from Ptolemy's Greek through British, Anglo-Saxon, Old English, Welsh, Latin and French forms. Ekwall gives us Tamesa, Tamesis, Tamensis, Tamessa, Tamessis, Tamis, Temys, Temes, Temese, Taemis, and several others. There is a prodigious volume of learning on display. But when it comes down to the root, the first syllable, Ekwall has this to say: 'Holder suggests that the name

Thames is related to Sanskrit *Tamasa*, the name of a tributary of the Ganges, whose meaning is "dark water". Whether Thames belongs to the root *teme-* or not, I think it very likely that the name Thames does. For this river the meaning "dark river" is very appropriate, at least for parts of it.' Really? Excuse me if I am being dense, but the Thames is no darker than many other flow of water, and why should the name of a river in India be picked up in ancient Britain?

With the Trent, Ekwall states that its base is the British *Trisanton* — or *Trisantona.* 'It is obvious,' he says reassuringly. He then breaks his compound down: *tri* meaning 'across', *santon* from Welsh *hynt*, Old Irish *set*, Goth *sinps*, Old English *sip,* 'perhaps Latin *sentio*' — all meaning 'road' or 'journey'. Thus Ekwall arrives at the full meaning: 'one who goes across.' Then he adds a synonym: 'trespasser', and asserts 'such a name would be applicable to rivers liable to floods', and invokes the Trent's tidal wave, known as the Aegir or Eagre. This strikes me as conjecture getting out of hand. Why 'trespasser'? And why should a flood be seen as a trespasser? And why should the Trent's floods be so much more remarkable than any other river's floods, at a time when all rivers flooded regularly? I was not convinced.

Then I stumbled upon a less celebrated name in toponomical circles, that of Dr Thomas Benjamin Franklin Grantham Eminson, of Scotter in Lincolnshire. In 1934 he published a book with the racy title of *The Place and River Names of the West Riding of Lindsey Lincolnshire*, much of it culled from two papers which had appeared many years before in the series *Lincolnshire Notes and Queries.*

It would be idle to pretend that Dr Eminson was a flowing stylist, or moved by any desire to amuse the reader. The papers are dense; there are ten thousand words about the Trent and its associated place names, which is a heap of toponomy. The curious thing is that his conclusions bear almost no relation to those of Professor Ekwall. It's almost as if they were studying two entirely different subjects.

Dr Eminson's point of departure is what he says is the 'general

British word' for river or water, which was *oy*, with variants *ay* and *ey*. He maintains that a river bend, or junction with a tributary, was *hum*, *hamm*, *hemm*, or *homm*; and that a landing place was *heth*. Much, in Dr Eminson's view, flowed from these roots. The Thames, for instance, was the *hamme eyse* – 'the winding waters'. The Derwent and Darenth were *de ay re hemm heth* – 'the landing place in the winding river reach'.

According to Dr Eminson, the name Trent is a contraction of 'Trenth' which itself is a contraction of the compound *the re hemm heth* – 'the landing place on the river bend'. He suggests that this was applied to one particular stretch, or independently to several different ones, and gradually spread along the whole river. It's a matter of guesswork how long it took for a single name, with local variants, to gain acceptance. But in time it had to be written down, and the Romans – says Dr Eminson – had difficulty with the *th-* sound and suppressed it in favour of a hard *t-* (in the same way Caesar wrote Cantuum for Kent, instead of the Anglo-British Canth or Cuenth). Bede recorded the river as Treant or Treenta; subsequent Latinised versions include Treonta and Trenta as well as Trisantona. But Dr Eminson believed that, in the spoken language, it held on to its ancient British form as Trenth, Trensh, Trench, Treonth, Treanth, or Trenith long after the Romans had gone and the Normans had been absorbed. As late as 1599 a scribe employed to record the proceedings of the Michaelmas court at Scotter – which is on the east side of the river north of Gainsborough – used both the official form, Trent, and a dialect variant, Trenth.

Eminson saw the influence of the river as paramount. Thus Nottingham comes from Snotingaham, which is squeezed together from *se homm heth inga hamm* – 'the enclosure of grass lands at the landing place near the river'. Walkerith, just downstream from Gainsborough, draws together *heth* and *wealcan*, meaning 'roll from side to side' – a reference, asserts Dr Eminson, to the Aegir, giving the meaning 'the landing place at the rolling reach'. His own

Scotter, three miles from the river, comes from *se cote re* — 'the dwelling near the river reach'; he claims that it was accessible from the Trent up its tributary, the Eau. Frodingham, on the same ridge as Scotter, was also accessible by river, in Dr Eminson's view — hence its name, derived from *the rey re oy heth ing hamm* — 'the grass land enclosure near the landing stage of the river reach'.

I like Dr Eminson's derivations. There is drama in them, and poetry and romance. They conjure a picture of a people engaged in a struggle with an unruly environment, searching for sanctuary from floods and rolling tides, fixing their landing places and enclosing their grass lands. In contrast, the origins offered by the English Place-Name Society — which was founded by Eilert Ekwall and is still going strong at the University of Nottingham — seem both mundane and incomplete: Nottingham — 'homestead of Snot's people'; Walkerith — 'fuller's landing place'; Scotter — 'Scot's tree'; Frodingham — 'homestead of Froda's people'. Who was Snot? Who was Froda? Why fuller? Why a Scot's tree?

Which authority should we believe, the great Ekwall or the forgotten Eminson? I regret to say that Spittal and Field, in their *Reader's Guide to the Place-names of the United Kingdom*, are sceptical about the Englishman.[3] They say Eminson's etymologies 'should be treated with caution', which I suspect is tantamount, in this gentlemanly field of scholarship, to classifying him as a crackpot. Even so, I still like to think that the sage of Scotter, looking down on the great river, may have had a feel for the landscape and its early inhabitants denied to the cerebral Swede. The subject is, for sure, a rich and absorbing one once you get into it, and I could go on. But perhaps not.

From Dunham a path took me through the fields and along the flood bank to Laneham, where there is a lovely little church on a knoll above the river. It used to be the estate of the Markhams, and inside the church, kneeling side by side, are two of them, father and son. The senior, Ellis, was regarded with great affection by

Queen Elizabeth. She included him in a pithy couplet about Nottinghamshire squires – 'Gervase the gentle, Stanhope the stout/Markham the Lion, and Sutton the lout' (the gentle Gervase being Clifton of Clifton Hall). His epitaph records that 'he long served her Majestie in her warres with extraordinary proofe in Ireland and the Low Countries'.

The son, Gervase, had a more chequered career. As a young man he made an enemy of another county notable, the rich and famously quarrelsome John Holles of Haughton. They were lawless times in the shires: Markham was attached to the interest of the Earl of Shrewsbury, and Holles to that of the Stanhopes, and the two families were at war. Trouble flared when a Shrewsbury man insulted a Stanhope retainer (by putting his thumb in his mouth and rattling his nail against his teeth) and was slain. Holles helped hush up the killing and was publicly accused of so doing by Markham, whereupon he charged Markham thus: 'He lyes most falsly . . . he lyes in his throte . . . when he dare (though otherwise he is a rascall and a vagabond and having nothing and I a thousand pounds a year) . . . do I trebble the ly uppon him'.[4]

They met one night by accident in Sherwood Forest, and rapiers were drawn. According to an account derived from the Holles family archive (possibly lacking objectivity), Markham 'rather capers and flourishes than fences' in front of Holles, 'supple as an eel, alert as a serpent and with a sting in him too'. Within moments Markham was indeed stung – 'pierced and spitted through the lower abdominal regions'. For a time his life hung in the balance, and all-out conflict between the feudal overlords threatened. But he pulled through, and promised 'never to eat supper nor to take the sacrament' until he was avenged. It proved to be a rash vow. Holles went on to become a considerable swell at court, while Markham stayed at home by the Trent. He lived another forty years, supperless and sacramentless, and the opportunity to get his own back never came.

The countryside north of Laneham is blighted by another power station, Cottam. The river executes two big loops to make its escape to the east. I was pedalling laboriously along the flood bank when, out of the mist clinging to the far bank, rose a line of toothy stumps of masonry. They were the ruins of Torksey Castle, which wasn't a castle but a fortified manor house and was sacked during the Civil War and has been rotting away ever since.

Nothing about Torksey today suggests that if you had happened this way a thousand years ago you would have found a bustling river port boasting two religious institutions, three churches, an array of merchants' houses and wharves lined with vessels loading and unloading lead, iron, corn, timber, wine, wool, dried fish, cured meat and all the other staples and good things of life.[5] Torksey owed its success to its position on the junction between two of the most important trading routes in eastern England: the Trent, and the artificial waterway known as the Fossdyke, dug by the Romans to provide a link between the Trent and the Witham via the regional capital, Lincoln. Torksey managed the trade both ways and charged tolls on both waterways, and for a time it flourished. But the Fossdyke was a great labour and expense to maintain. The merchants were forever complaining about it being silted up, commissions appointed by the King were forever ordering remedial work, while Torksey went downhill.

In 1530 Henry VIII's antiquary, John Leland, inspected it and reported that there was almost nothing left of the old buildings except ruins and earthworks. The Bishop of Lincoln offered a remission of sins to anyone who would repair the Fossdyke, blaming the 'indigence and disaster' that had overtaken the city on the canal's neglect. But the Bishop died and nothing much was done, and when the self-styled Water Poet, John Taylor, came this way a century later, he found its condition worse than ever.

Taylor was not a poet at all in the dictionary sense, rather an irrepressible fountain of ballads, lampoons, doggerel, squibs, satires, pamphlets, burlesques and mock-heroic verse; somewhere between Ogden Nash and Pam Ayres. At various times a Thames waterman, a bottleman (a kind of customs officer charged with collecting the duty on imported wines) and an impresario specialising in aquatic pageants, Taylor achieved celebrity in the role of boating buffoon. He was an energetic publicist for a succession of eccentric watery journeys, a fairly typical one from London to Queenborough in Kent in a craft made of brown paper propelled by oars made from dried fish tied to canes.

His *Verry Merry Wherry-Ferry Voyage* (you can see how a little of Taylor might go some way) took him and four cronies from London around the coast of East Anglia, via Cromer – where they were arrested on suspicion of being pirates – across the Wash to Boston and up the Witham to Lincoln.

> From thence we past a ditch of weedes and mud,
> Which they doe (falsely) there call Forcedyke Flood:
> For I'll be sworne, no flood could I finde there,
> But dirt and filth which scarce my boate would beare,
> 'Tis eight miles long and there our paines was such,
> As all our travell did not seem so much,
> My men did wade and drawe the boate like horses,
> And scarce could tugge her in with all our forces:
> Moyl'd, toyl'd, myr'd, tyr'd, still lab'ring ever doing,
> Yet were we nine long hours that eight miles going,
> At last when as the day was well nigh spent,
> We got from Forcedike's flood to Trent.[6]

Fortunately for the little band, after all their moyling, toyling and myrring, 'Trent's swift stream' soon had them in Hull where they were received as celebrities. They were feted and feasted by the mayor and aldermen, and Taylor reciprocated by portraying the port as a paradise on earth.

Underneath the clowning there was a serious point to Taylor. He was truly a creature of the liquid element, with a consuming passion for it and its possibilities. His mission was to proclaim the worth of water:

> Of famous rivers, bournes, rills and springs,
> Of deepes and shallows my invitation sings,
> Of rocks impenetrable, of fords and mills,
> Of stops and weirs, shelves, sands and mighty brills,
> Of navigable passages neglected
> Of rivers, spoiled, then beggared and directed.'

Like James Brindley later, John Taylor had a vision of a great network of waterways, feeding wealth through the body of the nation. Unfortunately for him, he lived through the wrong kind of changing times. During the Civil War he attached himself to the King's court-in-exile in Oxford where he supervised boat traffic on the river and assumed the part of court jester – the diarist Anthony Wood noted that 'he was much esteemed for his facetious company'. Later Taylor returned to London and set up as a publican, but irritated the Puritans by renaming his hostelry the Mourning Crown after the execution of Charles I. He then decided to call it the Poet's Head, and hung a picture of himself with the lines:

> There's many a head stands for a sign
> Then, gentle Reader, why not mine?
> Though I deserve not, I desire
> The laurel wreath, the poet's hire.

Alas, no one wanted to hire Taylor any more, and he sank into destitution, his death in 1653 being attributed to starvation.

I felt a touch sombre as I huffed and puffed along the path north from Torksey. Cottam's cooling towers steamed soundlessly to my left. Pylons tramped robotically through the drizzle in all directions. The Trent itself was stained terracotta with the silt stirred by the tide. It was ebbing fast and the margin of greasy, pockmarked mud along each bank was widening by the minute.

Littleborough Church

At Littleborough there was another church, older and even smaller than the one at Laneham. A barge heaped with gravel was moored mid stream, waiting for the tide to turn. The Romans

again

Waiting for passage

called this place Segelocum, and they paved the ford with slabs of stone held in place by oak piles. Harold of England marched his army across the river here on his way to meet William of Normandy at Hastings, and the Conqueror came in the opposite direction intent on applying some Norman discipline to the unruly north. These days it's a peaceful spot. The rooks and pigeons circled undisturbed around the open belfry, stooping when the mood took them to help themselves in the fields of leeks, kale and spinach.

I took the Roman road west away from Littleborough and the river. The first village I came to was Sturton-Le-Steeple, whose church has an enormous square tower bristling with pinnacles. Inside, so *Highways and Byways* informed me, is a marble slab commemorating the extraordinary gallantry of Sir Francis Thornhaugh, whose death fighting the Scots at the Battle of Preston in 1648 so enraged his regiment that they slaughtered every enemy in sight on the grounds that 'the whole kingdom of Scotland was too mean a sacrifice for that brave man'.[8]

The village seems cast down by its closeness to yet another power station, West Burton, and I wasn't in the mood for another church. I was, however, saddle-sore, thirsty and famished, and there was a pub. Inside, good cheer was in short supply. I had the impression that there might have been a woman's touch around the place once, but not any more. The landlord growled that the best he could do by way of lunch was a chip butty. After bringing it to me, he embarked on a lengthy grouse about trade. It was something to do with lay-offs at the power station or changes in shifts; I couldn't really be bothered to work out which. I told him that the landlord of the White Hart in Newton had boasted of turning over five grand a week. He grunted savagely and disappeared behind the bar.

West Burton power station

Some time in the afternoon, after a long, boring ride north, I reached the small town of Epworth, whose claim to fame is that it was the birthplace of John and Charles Wesley and thus the cradle of Methodism. My backside was aching like hell, so I stopped for a tour of the handsome Georgian rectory where the Wesley father, the Reverend Samuel, had the living. It was conducted by a plump, round-cheeked lady with curled grey hair and a smile that never dimmed for a moment. She spoke fluently about the extraordinary virtues of the Wesley boys, shuttling me briskly through a succession of rooms filled with items of furniture and knick-knacks that could have been there when the Wesleys were, but mostly weren't. We stopped for some time in front of a print on the stairs depicting the six-year-old John's rescue from a fire at the rectory. In the foreground the Reverend Samuel is shown on his knees, hands upraised towards heaven. 'An illustration of the power of prayer,' my guide said reverently.

Back on my bicycle I felt a sharp stab of regret, to go with the sharp stabs of pain in my buttocks, that I had ever thought of leaving the river so far behind. I still had another five miles to go, and I seemed to have been cycling for an eternity already. But it was too late to go back, so I went on until I got there. It was a place where a plan had been hatched long ago that literally changed the face of this part of England, and where a heavy price had been paid for the presumption.

My long ride had taken me across one of the least known and – the truth must be told – dullest tracts of countryside in England. At least on the other side of the Trent there is the occasional bulge of rising ground to vary the view, but to the west of the river it is just flat, flat, flat; with only the cooling towers and the pylons to intrude some verticality.

Such dullness can be useful, and this is, as a powerhouse of agricultural production. Potatoes, parsnips, beetroot, leeks, spinach, cereals, lawn turf – all grow here in easy abundance. The landscape is intersected by long, narrow, straight tracks along which, at the times dictated by the seasons, colossal trailers piled high with earth-encrusted root vegetables are dragged by gasping tractors. The rest of the time these fields sleep, disturbed only by the occasional visit from the muck-spreader or the pesticide and herbicide sprayer. The farms are far apart, brick houses bracketed between and dwarfed by gleaming barns and storage units. Everything is quiet and regimented. Yet this was once one of the wild, ungovernable badlands of England.

It was known (it still is) as the Isle of Axholme, the river island.[9] It was bound to the east by the Trent, and to the north, west and south by a ring of other tortuous, sluggish, murky rivers. The Isle was not an island, nor was it land. It was criss-crossed by reed-lined channels, ditches, drains and dykes, between which extended marshes, swamps, peat bogs, meres and morasses, broken here and there by a hummock of dryish ground rising a few feet above the quaking surrounds.

The Isle nurtured a watery way of life. The Islanders trapped eels and other fish, helped themselves to the wild ducks and geese and game they could lay their hands on. In summer, when the water receded, they grazed cattle on the lush pastures, and could grow almost any crop they wanted on the silt-rich soil. The reeds and woods provided them with fuel and building materials, and the waterways enabled them to come and go as they pleased. They kept themselves to themselves and regarded outsiders with the deepest suspicion. Their world was self-contained. They neither knew nor wanted any other.

Their misfortune was that their world-view became unfashionable once the drift towards centralised administration and organised economic development began to take hold. The Axholme economy was founded on rights of commons established in writing early in the fourteenth century and observed in practice long before that. With every cottage and its patch of land came commoners' rights: to graze the common land and to take fuel from it. Everyone had a stake in it, but no one – least of all the land-grabbing Crown – could claim to own it.

That was the Isle of Axholme. Immediately to the west of it, beyond the snaking River Don, stretched another forbiddingly damp and flat expanse of bog, marsh, moorland and water. Geographically, Hatfield Chase was indistinguishable from Axholme, and it supported a very similar way of life. The crucial difference was that it was royal property and its people were tenants of the King. The deer they took, the swans they roasted, the fish and fowl they trapped, the grazing land they turned their animals on to, their dank hovels – all were the King's. They were poachers, unruly, disrespectful, worst of all, unprofitable.

Land drainage – the transformation of marshland into a landscape fit for human use – was one of the great causes of Europe's emerging nation states. New worlds were being discovered by the explorers across the seas, and new worlds were being created at home. The making of them depended on new skills learned by a different breed: engineers, architects, map-drawers, technicians. These were practical men, proud of their crafts, more likely to be associated with merchants and entrepreneurs than with aristocratic patrons or centres of learning. Their relationship with the landscape was wholly different to that of the old feudal aristocracy and peasantry. They didn't see it as a God-given condition you were born into and stuck with, but as a God-given opportunity to which reason and intelligence, combined with acquired technical skills, were to be applied.

Hydraulically, the Dutch led the way. As early as the ninth century, Teutonic Frisians had migrated from southern Sweden to the Low Countries and northern Germany, establishing themselves on humps of drier ground which they built up by digging and protected as best they could with earth defences against the sea. Later, the invention of the windmill in the Netherlands obviated the use of manpower or horsepower to lift the water. The celebrated Dutch engineer Jan Leeghwater drained Lake Beemster, north of Amsterdam, and turned it into seven thousand hectares of prime agricultural land, and installed fifty windmills to keep it that way. Such achievements aroused interest across Europe, and the Dutch engineers and their gangs of skilled workers – the Polderjungen, or Polder Boys – made themselves available for hire. If you wanted a marsh drained, you sent for the Dutch.

In England, drainage talk had been buzzing around royal ears for years. Queen Elizabeth had sought the advice of the most celebrated water engineer of an older generation, Humphrey Bradley (a Dutchman, despite his name), on draining the Fens. Bradley told her he could do it in six months at a cost of £5000, and estimated it would bring £40,000 a year into Crown coffers. It was a tempting prospect, but local opposition and distractions elsewhere sunk the scheme. Twenty years on, in 1606, another Fenland scheme was put before James I. Again, nothing came of it, but the King remained a theoretical enthusiast. He appointed a commission to investigate the state of Hatfield Chase and 'whether his Tenants have not forfeited his Favour of Commoning therein by building new Houses upon it, joysting Beasts upon it, cutting down the Trees thereof, and destroying his Game; likewise they were to consider about the Draining, Improvement and Disafforestation thereof'.

The commission was mainly comprised of local landowners who were not above a little beast joysting of their own. They took the parochial view. Yes, the King was being 'encroached upon, his Favours abused, his Chase joysted in etc'. But, no, 'considering how great the Levels were and continually deep with Water, how many Rivers run thereinto and such like, they did humbly conceive that it was impossible to drain and improve them'.

But at the same time, James was getting a different view, and one that he much preferred. It came from Cornelius Vermuyden, a member of one of a close-knit network of prosperous drainage families from Zeeland. Vermuyden was well known, having supervised the draining of parts of the Royal Park at Windsor, as well as building a sea wall around Canvey Island. In the event James dithered and died without deciding anything. Just over a year later, Charles I – young, urgent, and in need of cash to fund his ambitions – signed an agreement with Vermuyden to drain Hatfield Chase.

The terms were simple. Of the acreage reclaimed, two-thirds was to go to the Crown, and one-third to Vermuyden and his backers, mainly Dutch merchants and financiers. Of the costs, three-thirds would be born by the Dutch, and none by the Crown. From the King's point of view, it was a beautiful arrangement, so beautiful that he was moved – or his advisers were – to conceal one crucial factor from Vermuyden. The agreement covered both Hatfield Chase and the Isle of Axholme; indeed, in engineering terms it had to, since much of the surplus water would be taken through Axholme to the Trent. But while Hatfield was Crown property, Axholme was not. Charles was lord of the principal manors of Epworth and Haxey, and no more. In that concealment was the root of all the trouble.

Vermuyden's plan, in essence, was to drain the water eastwards via the rivers Idle and Torne to the Trent, and north via the Don to the Ouse. Each of the smaller rivers was drastically reshaped and redirected to try to ensure a continuous discharge. Historically, the Don had an eastern branch which flowed into the Trent near its mouth, and which Vermuyden blocked off entirely. A supporting network of drains was needed. In some cases existing ones were used, in others new channels were dug.

Vermuyden chose for his headquarters a soggy little spot on the River Idle called Sandtoft. Having diverted the river to the east, he imported ready-made frames from Holland and put up two hundred houses for his Dutch, Flemish, Walloon and Huguenot workers, and set them digging. Within eighteen months – by the end of 1627 – he announced that the Hatfield part of the project was done. He applied for the reclaimed lands to be doled out according to the agreement, and then pressed ahead with the Axholme works. But no one had thought of asking the Commoners there for their permission.

They lost no time in making their feelings plain. In August 1628 they attacked Vermuyden's men near Haxey, stoned them and beat them, and destroyed the embankment they were working on. Vermuyden ordered his overseers to carry weapons. There was another assault, and one of the fenmen was killed. By then Vermuyden was under growing financial pressure. In addition to securing his own land in Hatfield, he had agreed to pay the King £10,000 for his share, plus a sizeable rent. His backers were pressing him to complete the whole project and deliver the dividend they expected. The digging went on under the protection of government militiamen.

Problems multiplied. Petitions and complaints poured in from villages along the Don, where people found their lands being regularly inundated because the new, single channel was not big enough to take all the water it was supposed to. In the end a new cut – still called Dutch River – had to be made to the Ouse, eating up £20,000 of the theoretical profits. In Axholme, a state close to insurrection prevailed. The drainage works were persistently sabotaged and there were sporadic outbreaks of rioting. Sandtoft, with its windmills, its gabled houses, its chapel – with services conducted in Flemish or French – and its general unEnglishness, became the particular target of xenophobic resentment.

Vermuyden, who was a canny operator, extricated himself as soon as he could from the bog of litigation and violence into which the venture was sinking. In 1630 he was engaged by the Earl of Bedford to take charge of the draining of the Great Level of the Fens. He departed Sandtoft for marshes new, leaving his fellow investors – known as the Participants – and the wretched workers to face the music. For a while a wary, uneasy peace was observed, but the outbreak of hostilities between the King and his Parliament gave the Islanders the chance they'd been waiting for.

'About the month of June 1642,' recorded William Dugdale in his *History of Embanking and Draining*,

they arose in tumult, brake down the fenns and inclosure of 4000 acres, destroyed all the corn growing and demolished the houses. And about the beginning of February ensuing, they pulled up the gates of Snow Sewer which, by letting in the tides from the River Trent, soon drowned a great part of Hatfield Chase, divers persons standing there with muskets and saying they would stay till the whole level was drowned and the inhabitants were forced to swim away like ducks.

The Commoners were ordered by a parliamentary committee meeting in Lincoln to repair the damage 'but upon the order being brought, Thos Peacock and others defended the doors with muskets and refused to obey it'.

With the end of the war and the overthrow of the hated monarch, the Islanders were full of hope that their old rights would be restored. But they found that England's new rulers were no more favourably disposed towards their brand of sturdy independence than the King had been. Recourse to the law got them nowhere, so, in desperation, they turned to a man regarded by Cromwell and his associates as almost as big a menace as the King. The Leveller leader, John Lilburne – ever eager for a fight – agreed to help. At the same time a local lawyer, Daniel Noddel, organised what amounted to the ethnic cleansing of Sandtoft. He and his men invaded the village, drove out the settlers and attacked the chapel – according to one account, ripping out the seats and lead and burying carrion next to the altar. The houses, barns and crops were burned. 'This is our common,' the foreigners were told, 'and you shall come here noe more unless you bee stronger than wee.'

They weren't, and in the end they gave up and abandoned Sandtoft to the obstreperous locals. Some returned to Holland. Others, hearing that there was work to be had in the Fens, migrated to Thorney, near Peterborough, where they were welcomed and eventually absorbed into the host community. The Islanders profited little from their defiance. Their association with Lilburne did them no good at all, either before or after he was forced into exile. A parliamentary committee found in favour of the Participants, who appointed a singularly determined Roman Catholic, Nathaniel Reading, to act as their enforcer. He hired muscle against the Islanders, pursued them through the courts, and in the end outmanoeuvred everyone by the expedient of living to the age of one hundred. By then Sandtoft, its church, and its Dutchness had been pretty much obliterated from this flat, damp landscape.

No one, however, had thought to tell the author of the Lincolnshire volume in the *Highways and Byways* series, Willingham Franklin Rawnsley. Two hundred years after Nathaniel Reading was laid to rest, his book came out with a brief reference to Sandtoft 'where, it is said, that the village is still largely Dutch'.

These words had led me to picture a little settlement looking as if Rembrandt or Brueghel might have painted it: gables and crooked roofs, with a little church full of memorials to departed drain diggers. In my enthusiasm, I had failed to notice the ominous words 'it is said'.

In the course of my long ride I had already crossed several of the greater and lesser arteries of Vermuyden's scheme, including the old Bickersdyke and the 'new' Idle. A couple of miles north-west of Epworth I came to the River Torne, a long, straight strip of motionless brown water fringed by reeds. I left it behind, pedalled between more fields, then came to a crossroads. Here, according to my OS map, was Sandtoft.

The name lives on

I looked around. There was nothing here to suggest the land of dykes and clogs and tulips, no gabled houses, no church; just a smattering of nondescript dwellings, a cluster of new homes, and a pub which was shut. I felt a surge of resentment against Mr Rawnsley. What kind of shoddy work was this? Why hadn't he taken the trouble to mount his cycle and inspect Sandtoft for himself?

I stared at the map for a long time. I needed to be in Gainsborough that night. Sandtoft was in the top half of the map sheet, Gainsborough towards the bottom of the bottom half. That meant Gainsborough was a long way away.

Owston Ferry 1920s

I got back on the infernal machine and returned to Epworth. From there I cycled south-east to Owston Ferry, which is close to the Trent and was once the river port for Axholme. It is said to show a Dutch influence in its architecture, but I couldn't be bothered to look. Somehow, on the way out of Owston Ferry, I managed to take the wrong road, and instead of reaching the Trent, I found myself in a tract of country devoid of humans, houses and signposts. At length I crossed yet another dead-straight, muddy, reedy strip of water and spotted an angler with a long pole sitting beside it. We discussed eels, bream, roach and other placid lovers of quiet, murky waters. He told me that if I kept going, I would reach West Stockwith, where the Idle and the Chesterfield Canal join the Trent.

East Stockwith 1920s

I did as he said. I had read somewhere that, a century ago, West Stockwith had five busy boatyards and a population of five thousand; and that now it had no boatyards and a population of 250. I couldn't have cared less, nor did I have the strength to raise my head to look at its old Dutch houses. All I could do was to keep going. At 6.30 I crossed the Trent into Gainsborough. I looked at my speedometer. It said that I had covered 115 kilometres that day. Never before in my life had I exceeded twenty.

Chapter 21

A River's a River

There was one other taker for breakfast at the White Hart in Gainsborough next morning. He was a lean, stubbly-scalped Geordie who was due in Scunthorpe later to try to sell something to somebody. Over the flabby egg, pink sausage and fat-soaked fried bread, he supplied vignettes of his life. I gathered there was, or had been, a wife somewhere, maybe a child or two. He seemed more concerned at no longer having lurchers. There was nothing, he said, to beat a night lamping on the moors after someone else's game.

He'd seen me the night before in the soulless J. D. Wetherspoon pub towards the river. He asked me if I'd noticed another pub, on the other side of the street. I hadn't. He'd gone in there and got chatting to a couple of lasses. The boyfriend of one of them had turned up and there'd been words, perhaps a spot of bother. 'Normally I'd have brought one of 'em back here,' he said in a matter-of-fact fashion. 'Mebbe both. Never had any problem getting women, me. Got the patter, haven't I?' He might have been going through his CV at a job interview. 'But I couldn't see the point.' I asked why not. 'It's the hernia, isn't it? Had the op, now it's worse than before. Can't do it, can I? So what's the point?'

The White Hart had had a room, £25 a night bed and breakfast, for which I was properly grateful in my condition. It had a restaurant as well, the girl said, which was more good news. But no chef, she said, which was a shame. That's why I ended up in the J. D. Wetherspoon. They did food, although I wished afterwards they hadn't because the gravy slopped over my lamb shank had a very odd taste and was responsible, I was sure, for a sharp attack of diarrhoea around dawn. The beer, though, was fine and absurdly cheap, and I'd had four or five pints before hauling my aching limbs back into the White Hart and passing out.

They like their history at J. D. Wetherspoon. The pub was called the Sweyn Forkbeard in memory of the man who first made Gainsborough a place to take notice of. He was the son of the King of Denmark, so might, I suppose, have been called Sweyn the Dane, although Forkbeard is more imposing (his father was known as Harold Blue-tooth). Having sailed his longships across the North Sea and up the Humber and Trent in the summer of 1013 on his third and final invasion of England, Sweyn made Gainsborough his base for offensive operations. There he received the submission of the northern chiefs, and from there he marched south, laying lands waste, burning churches, slaughtering the menfolk and ravishing the women, until pretty much the whole kingdom was ready to acknowledge him as king.

Little good did it do him. Back in Gainsborough a few months later, he sickened and died. You can take your choice between two versions of his death. According to the Danish chroniclers, having repented of a life of violence, treachery, brutality and persecution of Christians, he summoned his favourite son, Cnut, urged him to rule well and wisely and to promote Christianity, and then expired peacefully. The English version is that he called an assembly at which he denounced his favourite hate figure among the English, the martyred king St Edmund, and mocked his reputation for sanctity; whereupon a ghostly Edmund appeared on horseback wearing full armour and ran him through with a spear. Sweyn, so the chronicler related with relish, died in physical and spiritual torment.

The features of Gainsborough that had appealed to the Danish invaders stood it in good stead in more peaceful times. It could be reached easily from the sea by using the tides, while being far enough inland to give access into the heart of manufacturing and farming England. With the silting up of the Fossdyke, it replaced Torksey as the chief port on the lower Trent. Corn, wool, alabaster and other exports were barged down to Gainsborough and then transferred on to bigger vessels for transit to Hull, which by the early fourteenth century had established itself as a thriving maritime port. The flow of imports came the other way.

In the 1590s the Willoughbys of Wollaton had a depot in Gainsborough to organise the shipment of the coal. Not long afterwards Sir William Hickman – a London merchant who had moved his trading base to Gainsborough and taken over the town's finest house, the Old Hall – set up shops in the town selling tobacco, soap, oils, vinegar, port, madeira, brandy, cider, currants, lead shot and other consumables. Trade continued to flourish. By the middle of the eighteenth century more than four thousand tons of cheese was being moved through Gainsborough each year, not to mention the steady stream of beer casks from Burton, shipments of lead from Wirksworth in Derbyshire, coal from Wollaton and elsewhere, pottery from Stoke, iron, yarn, canvas, millstones, nails, linen from anywhere and everywhere.[1]

The opening of Gainsborough's stately three-arched stone bridge in 1791 – the only permanent crossing below Newark – symbolised the town's success, and seemed to offer a solid guarantee for the future. Ten years later steam-powered barges were working up and down from Hull, cutting the journey time from two or three days to a matter of hours. In 1834 thirty thousand tons of coal, lime and stone was landed at Gainsborough for local consumption; and fifty thousand tons was shipped through. The town was regarded as sufficiently important in trading terms to be given its own customs facilities.

This is the Gainsborough of George Eliot's *The Mill on the Floss*, which she called St Oggs.[2] Outwardly respectable and godly, manifesting the full range of bourgeois virtues, within it seethes with malice, envy, hypocrisy and mean-spiritedness. Peopled by gossiping housewives, cheese-paring shopkeepers, pompous auctioneers, moralising widows, grasping financiers, self-satisfied wool staplers, bullying mill-owners and the odd haughty aristo, it is a place built on public morality and awash with private selfishness and suffocating righteousness.

In these respects it is like other provincial towns dissected in cold-eyed fashion by other Victorian novelists. The individuality of St Oggs lies in its geography, specifically its necessary relationship with water. George Eliot places the town in a landscape defined by one great river and its tributaries, intersected by dykes, drainage channels, canals and embankments, studded with sluice gates, flood arches, locks, wharves and mills. Its good fortune is its river, and the contrary forces that sweep it: 'A wide plain, where the broadening Floss hurries on between its green banks to the sea, and the loving tide, rushing to meet it, checks its passage with an impetuous embrace. On this mighty tide the black ships – laden with the fresh-scented fir planks, with rounded sacks of oil-bearing seed or with the dark glitter of coal – are borne along to the town . . .'

The town's strength is also its vulnerability. It appears to demonstrate the security of man's control over nature as well as the benefits that accrue, and the forward-looking burghers are forever pressing for yet more schemes to exploit and harness the water. But the miller, Mr Tulliver, has an intuitive understanding of the element denied to the money-grabbing planners, even if his lack of an 'eddication' makes it hard for him to put it into words. 'Water's a very particular thing,' he observes. 'You can't pick it up with a pitchfork . . . It's plain enough what's the rights and wrongs of water, if you look at it straightforrard; for a river's a river . . .'

Through the company of his own turning wheel, the miller is constantly reminded of something the townspeople have forgotten or never knew: that the power of the river cannot be taken for granted. Mr Tulliver recalls his father telling him 'as when the mill changes hands, the river's angry'. The novel's apocalyptic climax is engineered to show that, while nature can be subdued, it is never subjugated. The flood comes, the Floss becomes very angry indeed, punishing alike the guilty and the innocent – including the miller's daughter, Maggie.

George Eliot came to Gainsborough in 1859 to research the novel. By then the town had reached the highest point in its fortunes, and decline – gentle enough hardly to be noticed – had set in. River transport was cheap, but the railways were quicker, more reliable and getting cheaper. Railway track could be replaced when necessary, whereas the river had to be continually dredged to keep the channel clear. In 1844 Gainsborough collected more than £40,000 in customs dues; by 1880 this was down to £2000, and the following year the town lost its status as an independent port authority.[3]

Sailing days

Barge under sail

The Trent Navigation Company did its best to keep the river going as a trade route. The old family-run barges were gradually replaced by strings of unpowered containers dragged along by steam tugs. New locks were built to take bigger vessels. In 1933 the oil terminal at Colwick was opened, and the leading companies – including Shell and Esso – set up depots there. Other oil terminals came into operation at Newark and Torksey. Downriver from Gainsborough, at Flixborough, a new wharf was built with a rail link to the steelworks at Scunthorpe. Sand and gravel were excavated in

Paddlesteamer

enormous quantities up and down the Trent valley. As recently as the 1980s, 150-foot-long barges were in regular use as far upstream as Nottingham. But oil pipelines proved a lot cheaper and more convenient than barge transport, and the Colwick terminal closed, followed by the others. The collapse of the coal-mining industry inflicted another mortal blow to commercial river traffic.

Gainsborough pre-1914

Gainsborough went steadily downhill, but at least it remained a recognisable river port. Even twenty years ago, the wharves and warehouses were doing business, and barges and snub-nosed coasters jockeyed for position along the river front, loaded with grain, gravel and scrap metal. No more. The river life has departed, the warehouses and mills have been demolished or converted into flats.

The completion in 2000 of a new flood defence symbolised Gainsborough's severance from its past, the sheer wall acting as a frontier between the town and the abandoned Trent. The regular streets of drab red-brick terraces and the windswept wastes of the town centre carry no echo of the cobbled lanes, Georgian houses, bustling warehouses and factories of Miller Tulliver's day. The dislocation between past and present becomes more evident as you saunter along the Riverside Walk, with its hard benches, allocated spots of green space, interesting contemporary sculptures and historical panels. You lean over the black railings (approved by the Health and Safety Executive) and look down on the dark, silent flow and try to recreate the life and enterprise that once seethed here: the creak and groan of timbers, the crack of wind-filled canvas, the throaty growling of diesel engines, water churning, tillers thrust this way and that, hoists swinging, loads descending, the yelling, cursing, singing, protesting, encouraging in thick dialects and accents from across the land and the languages of a score of sea-faring nations. It's difficult.

Amid the rubbishy 1970s redevelopments, the mean, monotonous streets, the swirl of discarded litter and the clang of crumpled cans being booted around by bored teenagers, Gainsborough retains one startlingly splendid treasure, which is the Old Hall, where Sir William Hickman lived and where Sweyn probably had his terminal confrontation with St Edmund. I cycled very slowly around it, admiring its timbered exterior, massive buttresses, its tall chimneys and mullioned windows and canted bays. I didn't go inside to inspect its Great Hall, fearing that if I got off my bicycle I might never get back on it again. Then I took the road north.

Sloop, Morton Corner 1920s

289

It was mercifully flat and quiet, winding with the river's course between fields of sugar beet, grain and potatoes on one side and the rounded, grassy flood bank on the other. I hardly saw the Trent itself. When I did catch a glimpse, it was big and brown, its banks grey and slimy beneath the endless line of scrubby willow bushes. Almost without noticing them, I passed through Walkerith, East Stockwith and Wildsworth. The first bend downstream from Wildsworth, which looks like any other bend, is called Jenny Hurn and is the territory of a well-known fairy, Jinny (or Jenny) on Boggard. This minute sprite has a human body and the face of a seal, and is said to cross the river in a craft like a pie dish to feed on the available crops. I didn't see it, but I doubt if nine o'clock on a Thursday morning is the best time.

Ferry East Stockwith

These river margins were known as Ings, and although they were generally too flood-prone to be habitable, the regular depositing of silt meant they gave exceptionally rich grazing. They were also full of ancient trees buried in the peat which the locals dug out and dragged away for buildings and fuel. There were strict rules governing the management of the Ings and the authorities took a dim view of anyone who failed to fill in holes left after removing trees, as they weakened the flood defences and wandering cattle were liable to fall into them. In 1599 the Vicar of Messingham and Butterwick, the Reverend Richard Rowbottom, was fined twelve pence for failing to repair his excavations, and the court issued a warning that future offenders would have to pay forty shillings.

Keadby 1920s

Beyond East Butterwick I went under the flat, brown concrete bridge that takes the M18 east and west. Over to the right, out of sight, was Scunthorpe. Ahead, stuck across the river next to the

and again

Trent's last power station, Keadby, was a bridge made of latticed steel girders, looking like the work of a Meccano enthusiast. This is Keadby Bridge, which was built in 1916 and incorporated a swinging section to allow sailing barges to pass underneath without having to lower their masts. It hasn't swung for half a century, but it has its uses, one of which is to support the railway line between Scunthorpe and Doncaster.

For many miles I had been cycling in something of a trance, with only the occasional farm or hamlet to disturb the monotony of the landscape. But here the peace was abruptly fractured. Both Keadby on the west side of the river, and Gunness on the east, are industrial sites with their own wharves big enough to take real ships. Suddenly there were trucks grinding in all directions. One roared past me as I wobbled along the edge of Gunness Wharf, blowing my straw hat on to the verge.

The road cuts away from the river at Gunness, then rejoins it on the approach to the industrial estate at Flixborough, a name that still reverberates, faintly but insistently, across the decades since its moment in history. It came at 4.53 on a Saturday afternoon, 1 June 1974, when a temporary pipe carrying cyclohexane at a temperature of 300 degrees Fahrenheit fractured. Within a minute forty tons of the chemical — used in the manufacture of nylon — had leaked out and formed a cloud which ignited and exploded. The plant was destroyed and 1800 buildings over a mile radius were damaged. Twenty-seven people on the site were killed; had the explosion happened during the week rather than at the weekend, the entire workforce of more than five hundred would have perished.

The estate is now occupied by an assortment of less combustible manufacturing enterprises. Although it is less dangerous than it used to be, it is no more attractive to the passer-by, and as I panted past I began to feel distinctly disgruntled at the noise and ugliness I had run into. The road ascended quite steeply to the village of

View from Flixborough

Flixborough — steeply enough for me to have dismount and push. At the top I looked around me, and my spirits recovered. Below and behind me were the clutter of the wharves and the smoking sprawl of Scunthorpe. But if I turned my back on them and looked north and west, I saw green vistas of woods and meadows, with the smug and silver Trent swinging its way through the levels towards its end.

Chapter 22

Brown God

Flixborough is at the southern end of a ridge of chalk that extends north five miles or so almost to the Humber. Nowhere is the ridge more than 200 feet high, but in the context of the flatness around, it is like a mighty mountain. To the east it slopes gradually into the Lincolnshire Wolds, but its western flank drops steeply to the Trent's flood plain, conjuring tremendous views of the flatlands on either side of the river and its eventual union with the Ouse to form the Humber.

A path cut down from the outskirts of Flixborough to the plain and ran along the bottom edge of a thick wood that extended up the ridge and spilled over its crest. Somewhere over to my left the river was going about its business behind its flood banks. I cycled for a while, then walked for a while. Ahead the river curved back towards the ridge, meeting it at a point where it is sufficiently steep to be known as the Cliff.

Burton Stather 1920s

The settlement by the river, which amounts to nothing much at all these days, is called Burton Stather, while the main village, out of harm's way at the top, is Burton-upon-Stather. Staithe is the old word for landing place, and for many centuries this was the source of Burton's wellbeing. There was a ford here and later a ferry. A market was held here weekly, and this was the way goods and customers came in and out. Shipwrights set up business, and a pier was built in the 1860s as a stop for the steam packet service between Hull and Gainsborough. Much more recently, in the 1970s, a new wharf was built in one of the periodic and short-lived rushes of official enthusiasm for reviving inland water transport. The wharf is now defunct and when I trudged down there, the Ferry pub – where I had vague thoughts of finding a room – was shut and devoid of any sign of life.

Matters improved at the top of the hill. There was a fine, handsome pub, the Sheffield Arms, where there was a warm welcome, hearty food, first-rate beer and a bed – in fact everything the sorearsed traveller could hope for. The village, two long streets of old houses and cottages, was perfectly pleasant; and the Church of St Andrew promised what Pevsner referred to as 'an amazing N. arch' complete with octagonal abaci, multi-scalloped capitals, pellets and chamfered billets, whatever they might be.

In the normal course, I would have been happy to potter around Burton and give the bed some attention, but I had a journey to finish and its end was close. So after lunch I pushed my bicycle along the path past the church. It took me out of the village and along the top of the Cliff. Woods of sycamore, ash, oak, lime and chestnut gave way at intervals to grassy clearings from which I could look down on the snaking river. I skirted the grounds of a couple of fine old houses, Walcot Hall and Walcot Old Hall, though from a distance I couldn't make out which was which; then crossed a dusty, uncovered track that had been gouged into the slope and led down towards the river. Looking down, I glimpsed trucks heaped with spoil crawling through the flats, spewing dust clouds behind. Beyond, the tide was flowing and meeting a brisk downstream wind that churned the surface into short, choppy waves. A gravel barge cleaved upstream, and I shuddered to think about what might have befallen me and the *Otter*. 'All movements require very close correlation with the tide as well as expert pilotage,' cautions *Nicholson's Guide* at this point.

Cleaving upstream

I watched a buzzard quarter the sky. Pylons strode into the haze across land as flat as a table top. Butterflies hovered around the foxgloves. The path brought me to a fenced clearing overlooked by a bungalow. There was a lane, then more houses. I had reached Alkborough, the last village on the Trent. On the far side of the clearing, sunk into a saucer-shaped hollow in the grass, was a strange device: a maze cut in the turf, known – no one seems quite sure why – as Julian's Bower.

It is very old.[1] There is a description of it in the diary kept between 1683 and 1704 by Abraham de la Pryme, the historian of Hatfield, whose father was one of the Huguenot immigrants brought over to help Vermuyden dig his drains. De la Pryme refers

Maze

to it as 'Gillians Bore' and links it with another turf maze, long since erased, at Appleby to the south-east, describing them as 'nothing but great labyrinths cut upon the ground with a hill cast up round them for the spectators to sit about on to behold the sport'. It is likely that it had been around for at least four hundred years by the time de la Pryme inspected it, and that it was first cut by monks from a small Benedictine house nearby which was supported by the powerful monastery at Spalding. The friars may have intended it as an allegorical representation of the entanglement of sin out of which only the guiding hand of Providence can show the way; or simply as a place for fun and games.

With maze

Whatever its origins, it was a good place to sit and contemplate the river's end. I had followed it as a blue line on the map across England. Here, below me, looking like a fat snake coiled at rest in the sunshine, it finished.

Most rivers do not have that finality. The question of where they end hangs unanswered. Their mouths expand, stretching to suck in the flat lands either side. Land merges into water and sky, river into sea. There is no boundary, just the endless to-and-froing, back-and-forthing of the tides, land covered and uncovered. The estuary is its own world, made from river and sea but belonging to neither.

The Trent is different. Although twice a day its face changes as its direction is reversed, it does not surrender itself to the sea, but remains entirely and recognisably river. 'A river is within us, the sea is all around us,' Eliot says in *The Dry Salvages*. The Trent holds on to that self-hood until it meets the Ouse coming from the west. Then both are lost in the Humber, which has the name of a river but is, in fact, that other mingled entity, the estuary, until it gives way to the sea.

I was pleased that the journey should have so defined a finishing point. There had been times when I sighed for other, prettier, more romantic rivers and more picturesque places. But the Trent gave a sense of completion. It had a beginning, and I had stood beside it, and it had an end, which I was looking at. Now I needed to stand there as well.

From the far side

I freewheeled from the maze through Alkborough on to a farm track which led down and across the fields towards the water. The map showed a house, Flatts Farm, off to the right. But it was no longer there, and the only buildings left were a couple of barns. The track stopped short of the flood bank. As I propped up my bicycle beside a field of peas, a purple cloud of starlings rose from it and whirred inland. I clambered over the bank and across an expanse of bog grass until I was beside the water.

The place is called Trent Falls. According to Peter Lord, the river here 'joins its mate fiercely through dangerous rapids'. But there were no rapids, just the flood tide running hard into the face of the breeze and waves slapping against the muddy shoreline. Navigation lights winked red and green. A cormorant skimmed upstream like a flying spearhead, followed by a pair of flapping ducks. Sunshine glittered coldly on the bronzed surface. I felt like a very insignificant intruder beside something very old and powerful and remote. I thought again of Eliot's lines about the river as a strong brown god – 'sullen, untamed and intractable'. There were patches of blue sky overhead, but up the Ouse a storm was brewing, discharging mutters of thunder.

I stood there for a time with the wind in my face, then took out my notebook and attempted to find some arresting phrases to convey the scene and my feelings. Having failed utterly, I left and made my way back to the maze. Distance and height seemed to make the sight more manageable. A long, red, rusty barge nosed her way down the Trent towards the junction, so close to the near bank that she seemed to be creeping along the edge of the fields.

A squadron of earthmovers crawled along one of the tracks. They were engaged in a project organised by the Environment Agency to push back the flood defences and allow the Alkborough Flats to revert to something approaching their natural state. It was something to do with a strategy to 'manage the expected rise in sea levels resulting from global warming'. If all went to plan, this patchwork of vegetable and cereal crops would be transformed into a wild and quaking tract of saltmarsh, lagoon, mudbank, wetland and rustling reedbeds, the haunt of dunlin and bar-tailed godwit rather than pea-poaching starlings.

There was a stone slab near the maze supporting a coloured tableau called the Alkborough Millennium Plaque, showing the features of the panorama spread out below. On a clear day, it said, the pinnacles of York Minster were visible. I had to make do with more familiar shapes, the outlines of the cooling towers of the power stations at Ferrybridge and Eggborough, twenty-five and fifteen miles away respectively.

End – 2005

End – long ago

The sky in that direction was darkening all the time, and the thunder was now booming. Just in front of me, in the sunlight, swallows were banking and wheeling as they helped themselves to a hatch of some invisible insect. I thought again of the way I had come, the blue thread in the rough meadows below Biddulph Moor, the broad ribbon here. One and the same. In this end was its beginning.

Notes

Preface

1. Ted Hughes, *River* (Faber 1983).

2. Brian Waters, *Severn Tide* (Dent, 1947); *Severn Stream* (Dent, 1949).

3. Robert Gibbings, *Sweet Thames Run Softly* (Dent, 1940); *Coming Down the Wye* (Dent, 1942).

Chapter 1: 'I go on for ever'

1. R. S. Thomas, 'The River' from *The River's Voice*, an anthology of river poetry (Green Books, 2000).

2. This account is derived from E. C. Pielou, *Fresh Water* (University of Chicago Press, 1998).

3. Gaston Bachelard, *Water and Dreams: An Essay on the Imagination of Matter*, translated by Edith R. Farrell (Dallas: Pegasus Foundation, 1983).

4. Genesis, chapter 2, verses 10–14.

5. Walter Ralegh, *The History of the World*, part 1.

6. Virgil, *Aeneid*, book 8.

7. Revelation, 22: 1.

8. John Stewart Collis, *The Moving Waters* (Rupert Hart-Davis, 1955).

9. E. C. Pielou, *Fresh Water*, chapter 5, 'Flowing Water: Rivers and Streams.'

Chapter 2: Out of the Ground

1. Doctor John Brown quoted in Will Grant, *Tweeddale* (Oliver & Boyd, 1948).

2. Samuel Taylor Coleridge, 'Inscription for a Fountain on a Heath', Coleridge's Poems, edited by Ernest Hartley Coleridge, (Oxford 1945), page 381.

3. Alfred, Lord Tennyson, *Balin and Balan,* from *Idylls of the King* (lines 25–27).

4. Claudio Magris, *Danube* (Collins Harvill, 1990, pages 122–3).

5. Samuel Taylor Coleridge, *Biographia Literaria*, vol. 1 (Oxford University Press, 1979).

6. Samuel Purchas, *Purchas: His Pilgrimage*, book 4, chapter 13.

7. For sacred rivers see Simon Schama, *Landscape and Memory* (Harper-Collins, 1995), chapter 5, 'Streams of Consciousness'; Harry Brewster, *The River Gods of Greece: Myths and Mountain Waters in the Hellenic World* (I. B. Tauris, 1997); David Quint, *Origin and Originality in Renaissance Literature: Versions of the Source* (Yale University Press, 1983).

8. Ezekiel, 29: 10.

9. John Hanning Speke, from Chapter 4 of *What Led to The Discovery of the Source of the Nile*, John Hanning Speke (Blackwood 1864)

Chapter 3: Childish Things

1. Aristotle, *Metaphysics*, book 1.

2. Ovid, *Metamorphoses*, book 9.

3. E. C. Pielou, *Fresh Water*, chapter 6, 'Rivers at Work'.

4. Seneca, *Naturales Questiones*, book 3.

5. Philo Judaeus, *The Allegorical Commentary*, book 1.

6. Walter Ralegh, *The History of the World*, part 1.

7. Sir Philip Sidney, *A Work Concerning the Trueness of the Christian Religion*.

8. Ecclesiastes, 1: 7.

9. See Yi-Fu Tuan, *The Hydrological Cycle and the Wisdom of God: A Theme in Geoteleology* (University of Toronto Press, 1968).

10. Plato, *Phaedo*, paragraphs 600–624.

11. Aristotle, *Meteorologica*, book 1, part 13.

12. Bishop Isodore of Seville, *Etymologies*, chapter 14.

13. Pliny, *Natural History*, book 2, chapter 45.

14. Vitruvius, *De Architectura*, book 8.

15. Bernard Palissy, *The Admirable Discourses*, translated by Aurèle La Rocque (University of Illinois Press 1957).

16. Edmund Halley, *An Account of the Circulation of the watry Vapours of the Sea and of the Cause of Springs* (Philosophical Transactions of the Royal Society, Volume 16, pages 468–473).

17. John Wesley, *A Survey of the Wisdom of God in Creation*.

Chapter 4: The Magnificence of Cities

1. See Laurence W. Meynell, *James Brindley: The Pioneer of Canals* (Werner Laurie, 1956).

Chapter 5: North of South

1. W. G. Hoskins, *Midland England* (Batsford, 1949).
2. See Helen Jewell, *The North-South Divide* (Manchester University Press, 1994).
3. William Shakespeare, *Henry IV, Part One*, act 3, scene 1.
4. See Rob Smith, *Places in the Margin: Alternative Geographies of Modernity* (Routledge, 1991); and D. C. D. Pocock, *The Novelist's Image of the North* (Transactions of the Institute of British Geographers, new series, 1979).
5. George Orwell, *The Road to Wigan Pier*, chapter 7.
6. Interview with Lord Young quoted in David Smith, *North and South: Britain's Economic, Social and Political Divide* (Penguin, 1989).

Chapter 6: Holding Their Noses

1. Barry's designs are held at the Potteries Museum and Art Gallery, Bethesda Street, Stoke-on-Trent ST1 3DW.
2. Quoted in Denis Stuart, *Dear Duchess: Millicent Duchess of Sutherland* (Gollancz, 1982).

Chapter 7: Recreating the Spirits

1. Quoted in W. H. Herenden, *From Landscape to Literature: The River and the Myth of Geography*, Duguesne University Press, 1986.
2. T. S. Eliot, *The Dry Salvages*, stanza 1, lines 6–9, from *Four Quartets*.
3. Claudius Aelianus, *De Natura Animalium*, vol. 55.
4. Izaak Walton, *The Compleat Angler*, part 1, chapter 1.
5. Walton, *The Compleat Angler*, part 1, chapter 5.
6. Quoted in Volume 18 of the 1917 edition of the *Dictionary of National Biography* (Oxford University Press) in the entry for Dudley Ryder, 1st Earl of Harronby.

7. J. R. R. Tolkien, *Tale of Sun and Moon*, from *The Book of Lost Tales* (Allen & Unwin, 1983).

Chapter 8: Rocking Past Rugeley

1. E. C. Pielou, *Fresh Water*, chapter 6, 'Rivers at Work'.
2. See Lord Bagot, *Memorials of the Bagot Family* (1826).
3. See *The Journeys of Celia Fiennes* (MacDonald, 1983).

Chapter 9: The Ale of England

1. See Donald Worster, *Rivers of Empire: Water, Aridity and the Growth of the American West* (Oxford University Press, 1985).
2. See William White, *History, Gazeteer and Directory of Staffordshire* (1851).
3. Michael Drayton, *Poly-Olbion*, song 26.
4. See Introduction to *Water, Engineering and Landscape: Water Control and Landscape Transformation in the Modern Period*, edited by Geoff Petts and Denis E. Cosgrove (Belhaven Press, 1990).
5. Gaston Bachelard, *Water and Dreams*, pages 136–7.
6. See *A History of the County of Stafford*: Volume 9, *Burton-upon-Trent*, edited by Nigel J. Tringham (Boydell and Brewer, 2003); also accessible through www.british-history.ac.uk
7. See *Burton's Brewing Heritage*, www.burtoncamra.org.uk

Chapter 10: Crossing the Waters

1. Quoted in letter to Thomas Moore from Byron in Venice, dated March 31 1817, reproduced in Volume 3 of Moore's *Life of Lord Byron, with his Letters and Journals* (John Murray 1854).
2. Charles Cotton, *The Retirement: Irregular Stanzas Addressed to Mr Izaak Walton*, included in John Major's edition of *The Compleat Angler* (1823).
3. Daniel Defoe, *A Tour through the whole Island of Great Britain*, letter 6, part 2.
4. Quoted in Ronald Russell, *Rivers* (David & Charles, 1978).
5. See Ronnie H. Terpening, *Charon and the Crossing: Ancient, Medieval and Renaissance Transformations of a Myth* (Bucknell University Press, 1984).

6. John Milton, *Paradise Lost*, book 2, lines 575–80.

7. Referred to in Pausanias, *Description of Greece*, book 10.

8. Virgil, *Aeneid*, book 6.

9. Gaston Bachelard, *Water and Dreams*, chapter 3.

10. U. A. Fanthorpe, 'At Swarkestone', from *Collected Poems* (Peterloo Poets, 2005).

Chapter 11: The Jolly Miller

1. Philo of Byzantium, *Pneumatica* (Reichert, Wiesbaden) 1974.

2. Geoffrey Chaucer, *The Prologue* and *The Reeve's Tale* from *The Canterbury Tales*.

3. Alfred, Lord Tennyson, *The Miller's Daughter*, lines 99–102.

4. Edward Thomas, *The Mill-water* from *Last Poems*, (Selwyn and Blount 1918).

5. See Henry Blyth, *The Pocket Venus: A Victorian Scandal* (Weidenfeld & Nicolson, 1966), and A. J. Squires, *Donington Park and the Hastings Connection* (Kairos Press, 1996).

6. See J. M. Lees, *The Rise and Fall of a Market Town* (Transactions of the Leicestershire Archaeological and Historical Society, 1956).

Chapter 12: 'I constrained the mighty river'

1. The Honourable John Byng, *Torrington Diaries* (Eyre & Spottiswoode, 1934–8).

2. Karl Moritz, *Journeys of a German in England: A Walking Tour of England 1782* (Eland, 1983).

3. See A. C. Wood, *A History of Trade and Transport on the River Trent* (Transactions of the Thoroton Society, 1950), and Charles Hadfield, *Canals of the East Midlands* (David & Charles, 1966).

4. See Stephen Daniels, *Down in the Flood: George Eliot, D. H. Lawrence and the River Trent* from *Trentside*, edited by Nicholas Alfrey (Djanoogly Art Gallery, 2001).

5. Quoted in J. B. Firth, *Highways and Byways of Nottinghamshire* (Macmillan, 1924).

Chapter 13: Poets and Cannibals

1. Quoted in J. D. Chambers, *Victorian Nottingham* (Transactions of the Thoroton Society, 1959).

2. See A. C. Wood, *Sir Robert Clifton* (Transactions of the Thoroton Society, 1953).

3. George Gordon, Lord Byron from *The English Bards and Scottish Reviewers*.

4. Christopher Langman, *A True Account of the Voyage of the Nottingham Galley* (London, 1711).

Chapter 14: First Love

1. Mrs A. Gilbert (née Gee), *Recollections of Old Nottingham* (1904).

2. Another version of the story suggests that the argument was over the best way to hang game.

3. Byron, *Fragment, written shortly after the Marriage of Miss Chaworth*.

4. See Mrs L. Chaworth Musters, *The History of the Chaworth Family* (accessible through www.nottshistory.org.uk).

5. Byron, *The Dream*, verse II, lines 47–9.

Chapter 15: Poor Lame Boy

1. The account is given in Lucy Hutchinson, *Memoirs of the Life of Colonel Hutchinson* (Longman, Hurst, Rees and Orme, 1806). Col. Hutchinson took part in the engagement, but Mrs Hutchinson was a fiercely partisan narrator and not invariably reliable.

2. Fiona MacCarthy, *Byron: Life and Legend* (John Murray, 2002).

3. Edward Trelawney, *Recollections of Shelley, Byron and the Author* (Penguin, 1973).

4. Letter to John Hobhouse, see volume 6 of Byron's Letters and Journals edited by Leslie Marchand (John Murray, 1976).

5. See Roger Merryweather, *The Bramley: A World-Famous Cooking Apple* (Newark and Sherwood District Council, 1991).

Chapter 16: The Impostor

1. See David E. Roberts, *The Battle of Stoke Field* (Newark and Sherwood

District Council, 1987); Michael Bennett, *Lambert Simnel and the Battle of Stoke* (Sutton, 1987); Gordon Smith, *Lambert Simnel and the King from Dublin* (www.richardiii.net).

2. Francis Bacon, *History of the Reign of Henry VII and Selected Works* edited by Brian Vickers, (Cambridge 1998).
3. Borrowed from Gordon Smith's fascinating paper.
4. Eric W. Ives, *'Agaynst Taking Awaye of Women': The Inception and Operation of the Abduction Act 1487*, from *Wealth and Power in Tudor England: Essays Presented to S. T. Bindoff* (Athlone Press, 1978).

Chapter 17: Inns, Castles and Salmon Tails

1. See Bernard Smith, *The Ancient and Modern Trent*, from *Memorials of Old Nottinghamshire* (George Allen, 1912).
2. See A. C. Wood, *Nottinghamshire in the Civil War* (Clarendon Press, 1937).
3. Ascribed to Sir William Davenant, Poet Laureate, who died of syphilis in 1668.

Chapter 18: The Church by the Stream

1. See Brian Robinson, *Three Nottinghamshire Parishes: A Modern History of Averham, Kelham and Staythorpe.*
2. Mark Girouard, *Historic Houses of Britain* (Peerage Books, 1981).
3. Published Nottingham 1881, accessible through www.nottshistory.org.uk

Chapter 19: Redeeming Mr Sweetapple

1. See Smith, *The Ancient and Modern Trent*, from *Memorials of Old Nottinghamshire.*
2. F. H. West, *Sparrows of the Spirit* (S.P.C.K., 1961).
3. R. B. Outhwaite, *Sweetapple of Fledborough and Clandestine Marriage in the 18th Century* (Transactions of the Thoroton Society, 1990).

Chapter 20: The Mark of Cornelius Vermuyden

1. Michael Drayton, *Poly-Olbion*, song 26.
2. Eilert Ekwall, *English River Names* (Clarendon Press, 1928).

3. Jeremy Spittal and John Field, eds, *A Reader's Guide to the Place-Names of the United Kingdom* (Stamford: Paul Watkins, 1990).

4. See *Letters of John Holles* (published for the Thoroton Society, 1975).

5. See J. W. F. Hill, *Medieval Lincoln* (Cambridge University Press, 1948).

6. See *Travels Through Stuart Britain: The Adventures of John Taylor, Water Poet*, edited and selected by John Chandler (Sutton, 1999).

7. *John Taylor's Last Voyage*, 1641.

8. Lucy Hutchinson, *Memoirs of the Life of Colonel Hutchinson*.

9. See Vernon Cory, *Hatfield and Axholme: A Historical Review* (Providence, 1986); J. D. Hughes, *The Drainage Disputes in the Isle of Axholme* (Lincolnshire Historian, 1954); Clive Holmes, *17th Century Lincolnshire* (1980); L. E. Harris, *Vermuyden and the Fens* (Cleaver-Hulme Press, 1953); Trevor Bevis, *The River Makers*, 1991.

Chapter 21: A River's a River

1. See I. S. Beckwith, *Transport and Trade in the Lower Trent Valley in the 18th and 19th Centuries* (East Midlands Geographer, 1966).

2. See Stephen Daniels, *Down in the Flood: George Eliot, D. H. Lawrence and the River Trent.*

3. A. C. Wood, *The History of Trade and Transport on the River Trent.*

Chapter 22: Brown God

1. See W. H. Matthews, *Mazes and Labyrinths* (Longman, Green, 1922).

Index

Central Electricity Generating Board: commissions Rugeley Power Station, 99; commissions sculpture at Staythorpe, 219.

Chaplin, Henry, landowner and racehorse owner, 143–4.

Charles I: and Sir Gervase Clifton, 166; raises standard at Nottingham Castle, 180, 226; disgusted with Nottingham, 226; surrenders himself at Southwell, 228; orders Newark to surrender, 228; commissions Vermuyden to drain Hatfield Chase, 275–6.

Charles II, and Nottingham Castle, 180.

Charon, ferryman of the Styx: dismal figure, 125; not included by Homer, 126; included by Virgil, 126; not a pleasant sight, 127; boat, 127.

Charrington Brewery, Burton-upon-Trent, 114.

Chartley, Staffordshire, 80.

Chaucer, Geoffrey, poet: *The Reeve's Tale*, 48; *The Miller's Tale*, 138.

Chaworth family: at Annesley, 183; at Wiverton, 183.

Chaworth, Mary (Mrs John Musters): in Colwick Hall when attacked by rioters, 179–80; death, 181; heiress, 181; Byron falls in love with, 181–2, 186; and Mr Musters, 182; finds marriage disappointment, 187; pursues Byron, 187–8; assessment of own character, 188; buried, 190.

Chaworth, William, duel with 5th Lord Byron, 185.

Cheddar Gorge, visited by Coleridge and Southey, 18.

Chellaston, Derbyshire, 136.

Chia Jang, Taoist engineer, 99.

China as world's first hydraulic society, 99–100.

Chub, fish: at Mavesyn Ridware, 101; River Loddon, 117; Burton-upon-Trent, 118; Averham, 233; upstream from Dunham, 254.

Churches: St Wystan, Repton, 121; St Mary, Nottingham, 181; St Oswald,

East Stoke, 209; St Mary Magdalene, Newark, 224; St Michael and All Saints, Averham, 238; St Andrew, Burton-upon-Stather, 295.

Churnet, River, 104–5.

Civil War: Nottingham, 226–7; Newark, 227–9; Shelford Manor, 195.

Clifton Bridge, 171.

Clifton family, Clifton Hall: Sir Gervase, 166; Sir Robert, 166–7, 171.

Clifton Grove: in *Sons and Lovers*, 159; and Trent, 162; and Cliftons, 165, 167; glories, 167.

Clifton Grove, poem by Henry Kirke White, 170.

Clifton Village, 167.

'Clinton Arms', inn, Newark, 223.

Cnut, King of England, 283.

Coleridge, Samuel Taylor, poet: Robert Southey, 18, 19; ideals of Pantisocracy, 18–19; Culborne, 19; Ash Farm, 19; *Kubla Khan*, 19–20, 29–30; River Otter, 168.

Collis, John Stewart, writer: *The Moving Waters*, 13; mills, 141–2.

Colorado, River, 106.

Colwick Church, 190.

Colwick Hall: ransacked in Nottingham Riots, 179–80, 181; and Trent, 188; lake, 189.

Colwick Oil terminal, 287, 288.

Confucianism, 100.

Cook, John Parsons, alleged victim of William Palmer, 97.

Coors, American brewing firm, 114, 115.

Corbett, Bishop Richard: verses to Trent, 197; beauty of wife, according to Aubrey, 197.

Cottam Power Station, 266, 268.

Cotton, Charles, poet and angler: *The Compleat Angler*, 74, 120; Dove, 75, 120; friendship with Izaak Walton, 119–21; Beresford Hall, 119; as angler, 120; attacked by Sir John Hawkins, 120; builds Fishing-House, 120–1.

Cox, Richard, apple grower, 205.

Cox's Orange Pippin, apple, 205.